Algebra and Discrete Mathematics   Vol. 2

GRASSMANNIANS OF
CLASSICAL BUILDINGS

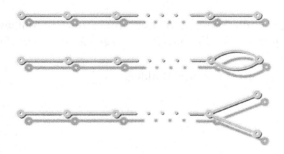

Algebra and Discrete Mathematics

ISSN: 1793-5873

Managing Editor: Rüdiger Göbel *(University Duisburg-Essen, Germany)*

Editorial Board: Elisabeth Bouscaren, Manfred Droste, Katsuya Eda, Emmanuel Dror Farjoun, Angus MacIntyre, H.Dugald Macpherson, José Antonio de la Peña, Luigi Salce, Mark Sapir, Lutz Strüngmann, Simon Thomas

The series ADM focuses on recent developments in all branches of algebra and topics closely connected. In particular, it emphasizes combinatorics, set theoretical methods, model theory and interplay between various fields, and their influence on algebra and more general discrete structures. The publications of this series are of special interest to researchers, post-doctorals and graduate students. It is the intention of the editors to support fascinating new activities of research and to spread the new developments to the entire mathematical community.

GRASSMANNIANS OF CLASSICAL BUILDINGS

Mark Pankov

University of Warmia and Mazury, Poland

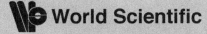 **World Scientific**

NEW JERSEY · LONDON · SINGAPORE · BEIJING · SHANGHAI · HONG KONG · TAIPEI · CHENNAI

Published by

World Scientific Publishing Co. Pte. Ltd.
5 Toh Tuck Link, Singapore 596224
USA office: 27 Warren Street, Suite 401-402, Hackensack, NJ 07601
UK office: 57 Shelton Street, Covent Garden, London WC2H 9HE

British Library Cataloguing-in-Publication Data
A catalogue record for this book is available from the British Library.

GRASSMANNIANS OF CLASSICAL BUILDINGS
Algebra and Discrete Mathematics — Vol. 2

Copyright © 2010 by World Scientific Publishing Co. Pte. Ltd.

ISBN-13 978-981-4317-56-6
ISBN-10 981-4317-56-X

Printed in Singapore.

Dedicated
to
my wife Inna
and
daughters Sabina and Varvara

Preface

Tits buildings or simple *buildings* are combinatorial constructions success-
fully exploited to study various types of groups (classical, simple algebraic,
Kac–Moody). One of historical backgrounds of this concept is Cartan's
well-known classification of simple Lie groups. We refer to [Abramenko
(1996); Brown (1989); Garrett (1997); Ronan (1989); Scharlau (1995); Tits
(1974)] for various aspects of building theory.

Buildings can be obtained from groups admitting Tits systems. Such
groups form a sufficiently wide class which contains classical groups, reductive algebraic groups and others. The formal definition of a building is pure
combinatorial and does not depend on a group. In [Tits (1974)] a building is defined as a simplicial complex with a family of subcomplexes called
apartments and satisfying certain axioms. All apartments are isomorphic
to the simplicial complex obtained from a Coxeter system which defines the
building type.

The vertex set of a building can be labeled by the nodes of the diagram
of the associated Coxeter system. The set of all vertices corresponding
to the same node is called a *Grassmannian* (more general objects were
investigated in [Pasini (1994)]). This term is motivated by the fact that
every building of type A_n is isomorphic to the flag complex of an $(n + 1)$-
dimensional vector space and the Grassmannians of the building can be
identified with the Grassmannians of this vector space. Every building
Grassmannian has a natural structure of a partial linear space (point-line
geometry); this partial linear space is called the *Grassmann space* associated
with the Grassmannian.

The aim of this book is to present both classical and more recent results on Grassmannians of buildings of classical types ($A_n, B_n = C_n, D_n$).
These results will be formulated in terms of point-line geometry. A large

portion of them is a part of the area known as *characterizations of geometrical transformations under mild hypotheses.* Roughly speaking, we want to show that some mappings of Grassmannians can be extended to mappings of the associated buildings. Other results are related with structural properties of apartments. Also we show that our methods work for some geometric constructions non-related with buildings — Grassmannians of infinity-dimensional vector spaces, the sets of conjugate linear involutions and Grassmannians of exchange spaces.

The book is self-contained and prospective audience includes researchers working in algebra, combinatorics and geometry, as well as, graduate and advanced undergraduate students. The requirement to the reader is knowledge of basics of algebra and graph theory.

Acknowledge

The present book contains the main results of my Doctor of Science (habilitation) thesis presented in Institute of Mathematics NASU and I thank A. Samoilenko and V. Sharko for supporting my research.

I am grateful to H. Van Maldeghem who read a preliminary version of the book and made a long list of valuable comments. Also I am grateful to H. Havlicek and M. Kwiatkowski for useful discussions and remarks.

Finally, I thank J. Kosiorek for drawing the pictures.

Mark Pankov

Contents

Chapter 0

Introduction

This short chapter is an informal description of the main objects of the book — buildings and their Grassmannians. All precise definitions will be given in Chapter 2.

The simplest buildings are so-called *Coxeter complexes* — a class of simplicial complexes defined by Coxeter systems. Let W be a group and S be a set of generators for W such that each element of S is an involution. The pair (W, S) is a *Coxeter system* if the group W has the following presentation

$$\langle\, S \,:\, (ss')^{m(s,s')} = 1,\ (s, s') \in S \times S,\ m(s, s') < \infty \,\rangle,$$

where $m(s, s')$ is the order of ss'. The associated Coxeter complex $\Sigma(W, S)$ is the simplicial complex whose simplices can be identified with special subsets of type $w\langle X \rangle$ with $X \subset S$ and $w \in W$. Every Coxeter system can be uniquely (up to an isomorphism) reconstructed from its diagram. The diagram associated with (W, S) is the graph whose vertex set is S and $s, s' \in S$ are connected by $m(s, s') - 2$ edges. In the case when W is finite, we get a Dynkin diagram without directions. There is a complete description of all finite Coxeter systems.

Similarly, more complicated buildings can be obtained from the *Tits systems*. A Tits system (G, B, N, S) is a structure on a group G consisting of two subgroups B, N which span G and a set of generators S of the quotient group

$$W := N/(B \cap N);$$

note that the pair (B, N) is called a BN-*pair*. By one of the basic properties of the Tits systems, (W, S) is a Coxeter system. The second remarkable property is the fact that every subgroup containing B, such subgroups are called *special*, can be reconstructed from elements of W and B. Moreover, there is a natural one-to-one correspondence between special subgroups and subgroups of W generated by subsets of S. The building associated with the Tits system (G, B, N, S) is the simplicial complex $\Delta(G, B, N, S)$ whose simplices can be identified with special subsets of type gP, where $g \in G$ and P is a special subgroup. If $w \in W$ then for every $g, g' \in N$ belonging to w and every special subgroup P we have $gP = g'P$; denote this special subset by wP. All such special subsets form a subcomplex Σ isomorphic to $\Sigma(W, S)$. The left action of the group G on the building defines the family of subcomplexes $g\Sigma$, $g \in G$. These subcomplexes are isomorphic to $\Sigma(W, S)$ and called *apartments*.

Following [Tits (1974)], we define an abstract *building* as a simplicial complex Δ with a family of subcomplexes called *apartments* and satisfying the following axioms:

- all apartments are Coxeter complexes,
- for any two simplices of Δ there is an apartment containing both of them,
- a technical condition concerning the existence of "nice" isomorphisms between apartments.

By this definition, every Coxeter complex is a building with a unique apartment. So, there exists a Coxeter system (W, S) such that all apartments of Δ are isomorphic to $\Sigma(W, S)$. The type of the building Δ is defined by the diagram of (W, S). We restrict ourselves to so-called *thick* buildings only; in such buildings maximal simplices form a sufficiently wide class. Coxeter complexes do not satisfy this condition.

Let V be an $(n+1)$-dimensional vector space and $\Delta(V)$ be the flag complex of V, i.e., the simplicial complex consisting of all flags of V. For every base B of V the subcomplex Σ_B which consists of all flags formed by linear subspaces spanned by subsets of B is called the *apartment* of $\Delta(V)$ associated with the base B. Every Σ_B is isomorphic to the simplicial complex of the Coxeter system (W, S), where W is the group of all permutations on the set $\{1, \ldots, n+1\}$ and S is formed by all transpositions $(i, i+1)$; the associated diagram is A_n. The simplicial complex $\Delta(V)$ together with the family of all such apartments is a building of type A_n. Note that this building can be obtained from the Tits system of the group $\mathrm{GL}(V)$.

Let Δ be a building and (W, S) be the associated Coxeter system. The vertex set of Δ can be naturally decomposed in $|S|$ disjoint subsets called *Grassmannians*. If the building is associated with a Tits system for a certain group G then the Grassmannians are the orbits of the left action of G on the vertex set. In the general case, this decomposition is related to the fact that the vertex set of Δ can be labeled by elements of S and this labeling is unique up to a permutation on S.

Let $\mathcal{G}$ be a Grassmannian of Δ. The intersection of $\mathcal{G}$ with an apartment of Δ is called an *apartment* of the Grassmannian $\mathcal{G}$. Two distinct elements $a, b \in \mathcal{G}$ are said to be *adjacent* if there exists a simplex $P \in \Delta$ such that $P \cup \{a\}$ and $P \cup \{b\}$ are maximal simplices. In this case, the subset formed by all $c \in \mathcal{G}$ such that $P \cup \{c\}$ is a maximal simplex will be called the *line* joining a and b. So, we get a structure known as a *partial linear space* or a *point-line geometry*, i.e., a set of points together with a family of subsets called lines and satisfying some simple axioms. This partial linear space is said to be the *Grassmann space* associated with $\mathcal{G}$.

The term "Grassmannian" is motivated by the following example. Let V be an $(n+1)$-dimensional vector space. The Grassmannians of the building $\Delta(V)$ are the usual Grassmannians $\mathcal{G}_k(V)$, $k \in \{1, \ldots, n\}$, formed by all k-dimensional linear subspaces of V. The Grassmann spaces corresponding to $\mathcal{G}_1(V)$ and $\mathcal{G}_n(V)$ are the projective space associated with V and the dual projective space, respectively. In particular, any two distinct elements of these Grassmannians are adjacent. The Grassmann space of $\mathcal{G}_k(V)$, $1 < k < n$, is more complicated. It contains non-adjacent elements. Note that our adjacency relation coincides with the classical adjacency relation introduced in [Chow (1949)]: two elements of $\mathcal{G}_k(V)$ are adjacent if and only if their intersection is $(k-1)$-dimensional.

Recall that a building is spherical if the associated Coxeter system is finite, and it is irreducible if the diagram is connected. *Irreducible thick spherical buildings* of rank ≥ 3 were classified in [Tits (1974)]. There are precisely the following seven types of such buildings:

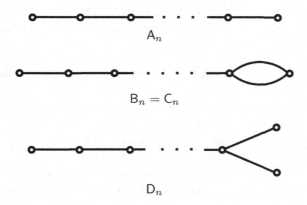

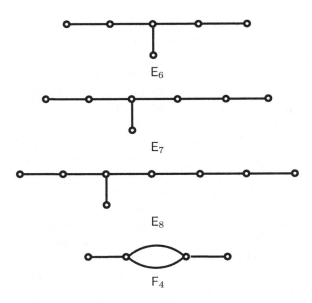

E_6

E_7

E_8

F_4

The first three types are called *classical*, the remaining four are known as *exceptional*. Every thick building of type A_n ($n \geq 3$) is isomorphic to the flag complex of a certain $(n + 1)$-dimensional vector space. All thick buildings of types C_n and D_n can be obtained from *polar spaces*, see Chapter 4. Exceptional buildings are related with so-called *metasymplectic* and *parapolar spaces*, see [Cohen (1995)].

In the present book we will consider Grassmannians associated with buildings of classical types only. Some information concerning Grassmannians of exceptional buildings can be found in [Cohen (1995)].

Investigation of building Grassmannians goes back to [Chow (1949); Dieudonné 2 (1951)] (see also Chapter III in [Dieudonné (1971)]) and is continued in [Cameron (1982)]. Currently, there are several research directions:

- axiomatic characterizations of Grassmann spaces, a survey can be found in [Cohen (1995)];
- embeddings in projective spaces, hyperplanes and generalized rank, see [Cooperstein (2003)] for a survey;
- subspaces of Grassmann spaces [Cooperstein, Kasikova and Shult (2005); Cooperstein (2005, 2007)];
- apartment properties [Blok and Brouwer (1998); Cooperstein and Shult (1997); Cooperstein, Kasikova and Shult (2005); Pankov 3 (2007)];

- characterizations of geometrical transformations of Grassmannians under "mild hypotheses" [Havlicek (1995); Huang (1998, 2000, 2001); Huang and Kreuzer (1995); Kreuzer (1998); Pankov, Prażmowski and Żynel (2006)] and the author's papers refereed in the book.

We describe all apartments preserving mappings of Grassmannians associated with buildings of classical types and collineations (isomorphisms) of the corresponding Grassmann spaces. Actually, the Fundamental Theorem of Projective Geometry and classical Chow's theorems [Chow (1949)] are partial cases of our results. The methods are based on deep structural properties of apartments (connections between the adjacency relation and apartments). Roughly speaking, we work in the latter two directions mentioned above. Also, we establish similar results for some geometric constructions non-related with buildings, for example, Grassmannians of infinite-dimensional vector spaces and the sets of conjugate linear involutions. One of them joins Chow's theorem with Dieudonné–Rickart's classification of automorphisms of the linear group [Dieudonné 1 (1951); Rickart (1950)].

Chapter 1

Linear Algebra and Projective Geometry

In Sections 1.1 and 1.2 we consider vector spaces over division rings and the associated projective spaces; it must be pointed out that vector spaces are not assumed to be finite-dimensional.

The main objects of the chapter are so-called semilinear mappings of vector spaces and the mappings of Grassmannians induced by them (Section 1.3).

By the Fundamental Theorem of Projective Geometry, every collineation (isomorphism) of projective spaces is induced by a semilinear isomorphism of the associated vector spaces. In Section 1.4 we prove a more general result known as Faure–Frölicher–Havlicek's version of the Fundamental Theorem.

The second result of Section 1.4 is Mackey's theorem concerning isomorphisms of the lattices formed by closed linear subspaces of normed vector spaces (real and complex). This theorem states that all such isomorphisms are induced by linear and conjugate-linear homeomorphisms of the associated normed spaces (the second possibility is realized only in the complex case).

The Fundamental Theorem of Projective Geometry can be reformulated in the following form [Baer (1952)]: every isomorphism of the lattices formed by linear subspaces of vector spaces is induced by a semilinear isomorphism. By this reason, Mackey's result is interpreted as the Fundamental Theorem for normed spaces.

In Section 1.5 we recall basic facts on reflexive sesquilinear forms.

1.1   Vector spaces

1.1.1   *Division rings*

Let R be a non-empty set with additive and multiplicative operations $+$ and $\cdot$ satisfying the following conditions:

- $(R, +)$ is an Abelian group (the identity element is denoted by 0),
- $(R \setminus \{0\}, \cdot)$ is a group (the identity element is denoted by 1).

If the distributive axioms

$$a(b + c) = ab + ac \ \ \text{and} \ \ (b + c)a = ba + ca \ \ \ \ \forall\, a, b, c \in R$$

hold then we say that $(R, +, \cdot)$ is a *division ring*. A division ring is called a *field* if the multiplicative operation is commutative.

Exercise 1.1. Show that $0a = a0 = 0$ and $-1a = -a$ for every $a \in R$.

For every division ring R we define the *opposite* (or *dual*) division ring R^* as follows: R and R^* have the same set of elements and the same additive operation, the multiplicative operation $*$ on R^* is defined by the formula

$$a * b = ba.$$

It is clear that R^{**} coincides with R, and we have $R^* = R$ only in the case when R is a field.

The following result is well-known; see, for example, Theorem 1.14 in [Artin (1957)].

Theorem 1.1 (J. H. M. Wedderburn). *Every finite division ring is a field.*

Let R and R' be division rings. We say that a mapping $\sigma : R \to R'$ is a *homomorphism* if

$$\sigma(a + b) = \sigma(a) + \sigma(b) \ \ \text{and} \ \ \sigma(ab) = \sigma(a)\sigma(b)$$

for all $a, b \in R$. Bijective homomorphisms are called *isomorphisms*.

Exercise 1.2. Show that the equalities

$$\sigma(0) = 0, \ \sigma(1) = 1, \ \sigma(-1) = -1$$

hold for every non-zero homomorphism $\sigma : R \to R'$.

Proposition 1.1. *Every non-zero homomorphism $\sigma : R \to R'$ is injective.*

Proof. If $\sigma : R \to R'$ is not injective then $\sigma(a) = 0$ for a certain non-zero element $a \in R$ and we have

$$\sigma(b) = \sigma(aa^{-1}b) = \sigma(a)\sigma(a^{-1}b) = 0$$

for every $b \in R$. □

An isomorphism of a division ring to itself is said to be an *automorphism*. The group formed by all automorphisms of R is denoted by $\mathrm{Aut}(R)$. An isomorphism of R to the opposite division ring R^* is called *anti-automorphism* of R (if R is a field then every anti-automorphism is an automorphism).

Proposition 1.2. *Every non-zero homomorphism of* $\mathbb{R}$ *to itself is identity. In particular,* $\mathrm{Aut}(\mathbb{R}) = \{1_{\mathbb{R}}\}$.

Proof. Let $\sigma : \mathbb{R} \to \mathbb{R}$ be a homomorphism. For every $n \in \mathbb{N}$ we have

$$n = \underbrace{1 + \cdots + 1}_{n},$$

hence

$$\sigma(n) = \underbrace{\sigma(1) + \cdots + \sigma(1)}_{n} = \underbrace{1 + \cdots + 1}_{n} = n.$$

This implies that $\sigma(m) = m$ for all $m \in \mathbb{Z}$. Since σ preserves division, the restriction of σ to $\mathbb{Q}$ is identity.

For every real number $a > 0$ there exists $b \in \mathbb{R}$ such that $a = b^2$, and we have

$$\sigma(a) = \sigma(b)^2 > 0.$$

This means that σ is order preserving: if $a > b$ then $a - b > 0$ and

$$\sigma(a) - \sigma(b) = \sigma(a - b) > 0$$

which implies that $\sigma(a) > \sigma(b)$.

Every $a \in \mathbb{R} \setminus \mathbb{Q}$ corresponds to unique section of rational numbers

$$\{\, q \in \mathbb{Q} : q < a \,\}, \quad \{\, r \in \mathbb{Q} : a < r \,\}.$$

Since σ is order preserving, it transfers this section in the section corresponding to $\sigma(a)$. On the other hand, σ sends every section to itself (its restriction to $\mathbb{Q}$ is identity) and we get $\sigma(a) = a$. □

The group $\mathrm{Aut}(\mathbb{C})$ is not trivial. It contains, for example, the complex conjugate mapping $z \to \bar{z}$.

Exercise 1.3. Show that every non-zero continuous homomorphism of $\mathbb{C}$ to itself is identity or coincides with the complex conjugate mapping. *Hint:* the restriction of every non-zero homomorphism $\sigma : \mathbb{C} \to \mathbb{C}$ to $\mathbb{Q}$ is identity and $\sigma(i) = \pm i$.

There are a lot of other automorphisms of $\mathbb{C}$, but they are non-continuous; moreover, there exist non-surjective homomorphisms of $\mathbb{C}$ to itself.

Example 1.1 (Finite fields). Every finite field coincides with a certain Galois' field $\mathrm{GF}(p, r)$ (p is a prime number). If $r = 1$ then we get $\mathbb{Z}_p$ and the group of automorphisms is trivial. In the general case, our field consists of p^r elements; its multiplicative group is a cyclic group of order $p^r - 1$; the group of automorphisms is a cyclic group of order r generated by the automorphism $a \rightarrow a^p$.

Example 1.2 (The division ring of real quaternions). Consider the division ring $\mathbb{H}$ formed by the real quaternion numbers $a + bi + cj + dk$. This is a 4-dimensional vector space over $\mathbb{R}$ (the canonical base consists of $1, i, j, k$) with the multiplicative operation defined by the following conditions

$$i^2 = j^2 = k^2 = -1,$$

$$ij = k, \quad jk = i, \quad ki = j.$$

This division ring is non-commutative (for example, $ik = i^2 j = -j = -ki$). Every automorphism of $\mathbb{H}$ is inner ($x \rightarrow qxq^{-1}$ for a certain $q \neq 0$). The conjugate mapping

$$a + bi + cj + dk \rightarrow a - bi - cj - dk$$

is an anti-automorphism.

1.1.2 *Vector spaces over division rings*

Let R be a division ring and $(V, +)$ be an Abelian group (the identity element is 0). Let also

$$R \times V \rightarrow V, \quad (a, x) \rightarrow ax$$

be a left action of R on V satisfying the following conditions:

(1) $1x = x$ for all $x \in V$,
(2) $a(bx) = (ab)x$ for all $a, b \in R$ and $x \in V$.

If the distributive axioms

$$a(x + y) = ax + ay \quad \text{and} \quad (a + b)x = ax + bx \quad \forall\, a, b \in R, \ x, y \in V$$

hold then we say that V is a *left vector space* over R. Elements of V and R are called *vectors* and *scalars*, respectively.

Similarly, every right action of R on V satisfying the conditions

(3) $x1 = x$ for all $x \in V$,

(4) $(xb)a = x(ba)$ for all $a, b \in R$ and $x \in V$

defines a *right vector space* over R if the corresponding distributive axioms hold. Clearly, this action can be considered as a left action, then (4) must be rewritten as

$$a(bx) = (ba)x \text{ or } a(bx) = (a * b)x,$$

where $*$ is the multiplicative operation of the opposite division ring.

Therefore, *every right vector space over R can be presented as a left vector space over R^*, and conversely.* In what follows we restrict ourselves to left vector spaces only.

Let V and V' be left vector spaces over a division ring R. A mapping $l : V \to V'$ is said to be *linear* if

$$l(x + y) = l(x) + l(y) \text{ and } l(ax) = al(x)$$

for all $x, y \in V$ and $a \in R$. We say that the vector spaces V and V' are *isomorphic* if there exists a linear bijection of V to V'. Linear bijections are called *linear isomorphisms*. The group of all linear automorphisms of V (linear isomorphisms of V to itself) is denoted by $\mathrm{GL}(V)$.

A subset $S \subset V$ is a *linear subspace* of V if

$$ax + by \in S$$

for all $x, y \in S$ and $a, b \in R$. By this definition, $\{0\}$ and V are linear subspaces. The intersection of any collection of linear subspaces is a linear subspace. For every linear mapping $l : V \to V'$ all vectors $x \in V$ satisfying $l(x) = 0$ form a linear subspace of V; it is called the *kernel* of l and denoted by $\mathrm{Ker}\, l$.

For every subset $X \subset V$ the intersection of all linear subspaces containing X is called the *linear subspace spanned* by X and denoted by $\langle X \rangle$; this subspace consists of all linear combinations

$$a_1 x_1 + \cdots + a_k x_k,$$

where $x_1, \ldots, x_k \in X$ and $a_1, \ldots, a_k \in R$. A subset $X \subset V$ is said to be *independent* if the linear subspace $\langle X \rangle$ is not spanned by a proper subset of X, in other words, for any distinct non-zero vectors $x_1, \ldots, x_k \in X$ the equality

$$a_1 x_1 + \cdots + a_k x_k = 0$$

holds only in the case when all scalars $a_1, \ldots, a_k$ are zero.

An independent subset $X \subset V$ is called a *base* of V if $\langle X \rangle$ coincides with V. Similarly, we define bases of linear subspaces. Every independent subset $X \subset V$ is a base of the linear subspace $\langle X \rangle$.

Proposition 1.3. *Every independent subset $X \subset V$ can be extended to a base of V; in particular, bases of V exist.*

Proof. Let $\mathfrak{X}$ be the set of all independent subsets containing X. This set is non-empty ($X \in \mathfrak{X}$) and it is partially ordered by the inclusion relation. If $\mathfrak{Y}$ is a linearly ordered subset of $\mathfrak{X}$ (for any $Y, Y' \in \mathfrak{Y}$ we have $Y \subset Y'$ or $Y' \subset Y$) then the subset

$$Z := \bigcup_{Y \in \mathfrak{Y}} Y$$

is independent. Indeed, suppose that

$$a_1 x_1 + \cdots + a_k x_k = 0$$

holds for non-zero vectors $x_1, \ldots, x_k \in Z$ and non-zero scalars $a_1, \ldots, a_k$; consider $Y_1, \ldots, Y_k \in \mathfrak{Y}$ containing $x_1, \ldots, x_k$, respectively; one of these subsets contains the others and is not independent, a contradiction. By Zorn lemma, the set $\mathfrak{X}$ has maximal elements. Every maximal element of $\mathfrak{X}$ is a base of V containing the subset X. $\square$

Theorem 1.2. *Any two bases of a vector space have the same cardinality (possible infinite).*

Proof. See Section II.2 in [Baer (1952)]. $\square$

The cardinality of bases of V is called the *dimension* of V and denoted by $\dim V$. Since every linear subspace can be considered as a vector space, the dimension of linear subspaces also is defined.

Every linear isomorphism maps independent subsets to independent subsets and bases to bases; thus isomorphic vector spaces have the same dimension. Every one-to-one correspondence between bases of V and V' can be uniquely extended to a linear isomorphism between these vector spaces.

Proposition 1.4. *For any two linear subspaces $S, U \subset V$ there exists a base of V such that S and U are spanned by subsets of this base.*

Proof. Suppose that $S \cap U \neq 0$. We take any base X of $S \cap U$ and extend it to bases B_S and B_U of S and U, respectively. Show that the subset $B_S \cup B_U$ is independent (this is trivial if one of the subspaces is contained in the other, and we consider the case when S and U are not incident).

If the statement fails then there exist distinct vectors

$$x_1, \ldots, x_k \in B_S, \ y_1, \ldots, y_m \in B_U$$

such that

$$\sum_{i=1}^{k} a_i x_i + \sum_{j=1}^{m} b_j y_j = 0$$

and all scalars a_i, b_j are non-zero. Then

$$\sum_{i=1}^{k} a_i x_i = -\sum_{j=1}^{m} b_j y_j \in S \cap U.$$

Since B_S and B_U both are independent, all x_i and y_j belong to $X = B_S \cap B_U$ which contradicts the fact that X is independent.

By Proposition 1.3, the subset $B_S \cup B_U$ can be extended to a base of V. This base is as required.

In the case when $S \cap U = 0$, the proof is similar. We leave it as an exercise for the reader. $\square$

For any two subsets $X, Y \subset V$ we define

$$X + Y := \{ x + y \ : \ x \in X, \ y \in Y \ \}.$$

If X, Y are linear subspaces then the sum $X + Y$ coincides with the linear subspace $\langle X, Y \rangle$.

Let S be a linear subspace of V. The associated *quotient vector space* V/S is formed by all subsets $x + S$ (we have $x + S = y + S$ if $x - y \in S$) and the vector space operations on V/S are defined as follows

$$(x + S) + (y + S) := (x + y) + S, \quad a(x + S) := ax + S.$$

A linear subspace U is called a *complement* of S if

$$S \cap U = 0 \quad \text{and} \quad S + U = V;$$

in this case, U is isomorphic to V/S (the mapping $x \to x + S$ is a linear isomorphism). Therefore, all complements of S have the same dimension called the *codimension* of S and denoted by $\operatorname{codim} S$.

Proposition 1.5. *Let S and U be linear subspaces of V. If these subspaces both are finite-dimensional then*

$$\dim(S + U) = \dim S + \dim U - \dim(S \cap U).$$

In the case when at least one of these subspaces is infinite-dimensional, we have

$$\dim(S + U) = \max\{ \dim S, \dim U \ \}.$$

This statement is a simple consequence of Proposition 1.4 and the following remark.

Remark 1.1. Recall that the cardinality of a set X is denoted by $|X|$. The sum of two cardinalities α and β is the cardinality $|X \cup Y|$, where X and Y are disjoint sets of cardinalities α and β (respectively). If $\alpha \leq \beta$ and β is infinite then $\alpha + \beta = \beta$.

For every linear subspace $S \subset V$ we have

$$\dim S + \operatorname{codim} S = \dim V;$$

in the infinite-dimensional case, this formula means that

$$\max\{\dim S, \operatorname{codim} S\} = \dim V.$$

Denote by $\mathcal{G}(V)$ the set of all linear subspaces of V. The Grassmannians of V can be defined as the orbits of the action of the group $\operatorname{GL}(V)$ on $\mathcal{G}(V)$; in what follows we do not consider two trivial Grassmannians formed by 0 and V, respectively. If $\dim V = n$ is finite then for every $k \in \{1, \ldots, n-1\}$ we write $\mathcal{G}_k(V)$ for the Grassmannian consisting of all k-dimensional linear subspaces of V. In the case when $\dim V = \alpha$ is infinite, there are the following three types of Grassmannians:

- $\beta < \alpha$ and $\mathcal{G}_\beta(V)$ consists of all linear subspaces of dimension β (the codimension of these linear subspaces is α),
- $\beta < \alpha$ and $\mathcal{G}^\beta(V)$ consists of all linear subspaces of codimension β (the dimension of these linear subspaces is α),
- $\mathcal{G}_\alpha(V) = \mathcal{G}^\alpha(V)$ consists of all linear subspaces whose dimension and codimension is α.

1.1.3 *Dual vector space*

Let V be a left vector space over a division ring R. Linear mappings of V to R are called *linear functionals*. We define the sum $v + w$ of linear functionals v and w as follows

$$(v + w)(x) := v(x) + w(x) \quad \forall\, x \in V;$$

and for any linear functional v and any scalar $a \in R$ the linear functional va is defined by the formula

$$(va)(x) := v(x)a \quad \forall\, x \in V.$$

The set of all linear functionals together with the operations defined above is a right vector space other R. The associated left vector space over R^* will be denoted by V^* and called the *dual vector space*.

Let $B = \{x_i\}_{i \in I}$ ($|I| = \dim V$) be a base of V. There is a one-to-one correspondence between linear functionals and mappings of the base B to R, since every linear functional can be uniquely reconstructed from its restriction to B. Denote by B^* the subset formed by the linear functionals x_i^*, $i \in I$, satisfying

$$x_i^*(x_j) = \delta_{ij}$$

(δ_{ij} is Kroneker symbol). This subset is independent.

If $\dim V = n < \infty$ then for every $x^* \in V^*$ we have

$$x^* = \sum_{i=1}^{n} x^*(x_i)x_i^*.$$

Thus B^* is a base of V^* and $\dim V = \dim V^*$. The base B^* will be called *dual* to B.

If V is infinite-dimensional then B^* is not a base of V^*. Indeed, a linear functional cannot be presented as a linear combination of elements from B^* if it takes non-zero values on an infinite subset of B. In this case, $\dim V < \dim V^*$ (see Section II.3 in [Baer (1952)] for the details).

Lemma 1.1. *The kernel of every non-zero linear functional is a linear subspace of codimension* 1.

Proof. Let $v \in V^* \setminus \{0\}$. If the codimension of $\operatorname{Ker} v$ is greater than 1 then consider $S \in \mathcal{G}_2(V)$ contained in a complement of $\operatorname{Ker} v$. There exist linearly independent vectors $x, y \in S$ satisfying $v(x) = v(y)$. Then $v(x - y) = 0$ which contradicts the fact that S intersects $\operatorname{Ker} v$ precisely in 0. $\qquad\square$

Exercise 1.4. Show that

$$\operatorname{codim}(S \cap U) \leq \operatorname{codim} U + 1$$

for every linear subspace $U \subset V$ and every linear subspace $S \subset V$ of codimension 1. As a consequence establish that

$$\operatorname{codim}(S_1 \cap \cdots \cap S_k) \leq k$$

for any linear subspaces $S_1, \ldots, S_k \subset V$ of codimension 1.

For every subset $X \subset V$ the linear subspace

$$X^0 := \{ \, v \in V^* \; : \; v(x) = 0 \;\; \forall \, x \in X \, \}$$

is known as the *annihilator* of X. Similarly, for every $Y \subset V^*$ the linear subspace

$$Y^0 := \{ \, y \in V \; : \; w(y) = 0 \;\; \forall \, w \in Y \, \}$$

is called the *annihilator* of Y. If S and U are linear subspaces of V or linear subspaces of V^* then $S \subset U$ implies that $U^0 \subset S^0$.

Proposition 1.6. *If $S \subset V$ is a linear subspace of finite codimension then* $\dim S^0 = \operatorname{codim} S$ *and* $S^{00} = S$. *Similarly, if U is a finite-dimensional linear subspace of V^* then* $\operatorname{codim} U^0 = \dim U$ *and* $U^{00} = U$.

Proof. Suppose that $\operatorname{codim} S = k < \infty$. Consider a base $B = \{x_i\}_{i \in I}$ of V such that S is spanned by a subset of B. Let B^* be the subset of V^* which consists of all linear functionals $x_i^*, i \in I$, satisfying $x_i^*(x_j) = \delta_{ij}$. There are precisely k distinct vectors from B which do not belong to S, denote them by $x_{i_1}, \ldots, x_{i_k}$. The restriction of $v \in S^0$ to B can take non-zero values only on these vectors. Therefore, $x_{i_1}^*, \ldots, x_{i_k}^*$ form a base of S^0 and $\dim S^0 = k$. Since S coincides with the intersection of the kernels of $x_{i_1}^*, \ldots, x_{i_k}^*$, we have $S^{00} = S$.

Let $\{v_1, \ldots, v_k\}$ be a base of U. Then U^0 is the intersection of the kernels of $v_1, \ldots, v_k$ and, by the second part of Exercise 1.4, its codimension is equal to $m \le k$. Since $\dim U^{00} = m$ (it was established above), the inclusion $U \subset U^{00}$ implies that $U^{00} = U$ and $m = k$. $\square$

Denote by $\mathcal{G}_{\mathrm{fin}}(V)$ and $\mathcal{G}^{\mathrm{fin}}(V)$ the sets of all proper linear subspaces of V with finite dimension and finite codimension, respectively. If V is finite-dimensional then $\mathcal{G}_{\mathrm{fin}}(V)$ and $\mathcal{G}^{\mathrm{fin}}(V)$ both coincide with $\mathcal{G}(V)$. Proposition 1.6 gives the following.

Corollary 1.1. *The annihilator mapping $S \to S^0$ is a bijection of $\mathcal{G}^{\mathrm{fin}}(V)$ to $\mathcal{G}_{\mathrm{fin}}(V^*)$. This bijection reverses inclusions:*

$$S \subset U \;\Longleftrightarrow\; U^0 \subset S^0.$$

If $\dim V = n < \infty$ then the annihilator mapping transfers $\mathcal{G}_k(V)$ to $\mathcal{G}_{n-k}(V^*)$ for every $k \in \{1, \ldots, n-1\}$. In the infinite-dimensional case, it sends $\mathcal{G}^k(V)$ to $\mathcal{G}_k(V^*)$ for every natural k.

Every vector $x \in V$ defines the linear functional $v \to v(x)$ of V^* (an element of V^{**}). If V is finite-dimensional then this correspondence is a linear isomorphism between V and V^{**}. In the finite-dimensional case, we will identify the second dual vector space V^{**} with V.

1.2   Projective spaces

In the first subsection we introduce the point-line geometry language which will be exploited throughout the book. Projective spaces over division rings will be considered in the second subsection.

1.2.1   *Linear and partial linear spaces*

Let P be a non-empty set and $\mathcal{L}$ be a set consisting of proper subsets of P. Elements of P and $\mathcal{L}$ will be called *points* and *lines*, respectively. Two or more points are said to be *collinear* if there is a line containing them. We say that the pair $\Pi = (P, \mathcal{L})$ is a *partial linear space* if the following axioms hold:

(1) each line contains at least two points and for every point there is a line containing it;

(2) for any two distinct collinear points p and q there is precisely one line containing them, this line will be denoted by $p\,q$.

A *linear space* is a partial linear space, where any two points are collinear. In what follows we will always suppose that a partial linear space contains more than one line.

We say that partial linear spaces $\Pi = (P, \mathcal{L})$ and $\Pi' = (P', \mathcal{L}')$ are *isomorphic* if there exists a bijection $f : P \to P'$ such that $f(\mathcal{L}) = \mathcal{L}'$; this bijection is called a *collineaton* of Π to Π'. A bijection of P to P' is said to be a *semicollineaton* of Π to Π' if it maps lines to subsets of lines. An injection of P to P' sending lines to subsets of lines is called an *embedding* of Π in Π' if distinct lines go to subsets of distinct lines.

Let $\Pi = (P, \mathcal{L})$ be a partial linear space. For every subset $X \subset P$ we define

$$\mathcal{L}_X := \{\, L \cap X \ : \ L \in \mathcal{L}, \ |L \cap X| \geq 2 \,\};$$

the pair $(X, \mathcal{L}_X)$ is called the *restriction* of the partial linear space Π to the subset X.

The *collinearity graph* of Π is the graph whose vertex set is P and whose edges are pairs of distinct collinear points. Our partial linear space is called *connected* if the collinearity graph is connected. In a connected graph the distance $d(v, w)$ between two vertices v and w is the smallest number i such that there is a path of length i connecting v and w; every path between v and w consisting of $d(v, w)$ edges is said to be a *geodesic*. If Π is connected then we define the *distance* between two points of Π as the distance between the corresponding vertices of the collinearity graph.

We say that $S \subset P$ is a *subspace* of Π if for any two distinct collinear points $p, q \in S$ the line pq is contained in S. By this definition, every set of mutually non-collinear points is a subspace. A subspace is said to be *singular* if any two points of this subspace are collinear (the empty set and one-point subspaces are singular). The intersection of any collection of subspaces is a subspace; moreover, it is a singular subspace if all subspaces from the collection are singular.

Exercise 1.5. Show that maximal singular subspaces exist and every singular subspace is contained in a certain maximal singular subspace. *Hint:* use Zorn lemma.

The minimal subspace containing a subset $X \subset P$ (the intersection of all subspaces containing X) will be called the *subspace spanned* by X and denoted by $\langle X \rangle$. In the general case, the subspace spanned by a clique of the collinearity graph (a subset consisting of mutually collinear points) does not need to be singular. We say that a subset $X \subset P$ is *independent* if the subspace $\langle X \rangle$ is not spanned by a proper subspace of X.

Let S be a subspace of Π (possible $S = P$). An independent subset $X \subset S$ is said to be a *base* of S if $\langle X \rangle = S$. We define the *dimension* of S as the smallest cardinality α such that S has a base of cardinality $\alpha + 1$ (if the cardinality α is infinite then $\alpha + 1 = \alpha$). The dimensions of the empty set and one-point subspaces are equal to -1 and 0 (respectively), lines are 1-dimensional subspaces. Two-dimensional linear spaces and two-dimensional singular subspaces of partial linear spaces are called *planes*.

A triple of mutually collinear points is said to be a *triangle* if these points are not collinear (in other words, the points span a plane).

A proper subspace of a partial linear space is called a *hyperplane* if it has a non-empty intersection with every line (it is clear that a hyperplane contains a line or intersects it precisely in a point).

Recall that in a *projective plane* each line contains at least three points and any two distinct lines have a non-empty intersection (by the second axiom of partial linear spaces, this intersection is a point). A *projective space* is a linear space where every line contains at least three points and the following axiom holds: if a, b, c, d are distinct points and the lines ab and cd have a non-empty intersection then the lines ac and bd have a non-empty intersection. Projective spaces can be also defined as linear spaces where all planes are projective.

1.2.2 *Projective spaces over division rings*

Let V be a left vector space over a division ring and $\dim V \geq 3$ (the dimension of V is not assumed to be finite). For every $S \in \mathcal{G}_2(V)$ the subset consisting of all 1-dimensional linear subspaces of S, in other words, $\mathcal{G}_1(S)$ is called a *line* of $\mathcal{G}_1(V)$. Denote by $\mathcal{L}_1(V)$ the set of all such lines and consider the pair

$$\Pi_V := (\mathcal{G}_1(V), \mathcal{L}_1(V)).$$

For any distinct $P_1, P_2 \in \mathcal{G}_1(V)$ there is the unique line $\mathcal{G}_1(P_1 + P_2)$ containing them. Every line $\mathcal{G}_1(S)$ contains at least three points (if vectors x, y form a base of S then $\langle x \rangle$, $\langle y \rangle$, $\langle x+y \rangle$ are three distinct points on the line). Thus Π_V is a linear space. In the case when $\dim V = 3$, any two lines have a non-empty intersection and Π_V is a projective plane.

If P_1, P_2, P_3, P_4 are distinct points of Π_V and $P_1 + P_2$ has a non-zero intersection with $P_3 + P_4$ then all P_i are contained in a certain 3-dimensional linear subspace; in particular, $P_1 + P_3$ and $P_2 + P_4$ have a non-zero intersection. Therefore, Π_V is a projective space.

Exercise 1.6. Show that $\mathcal{S} \subset \mathcal{G}_1(V)$ is a subspace of Π_V if and only if there exists a linear subspace $S \subset V$ such that $\mathcal{S} = \mathcal{G}_1(S)$.

Let $\mathcal{X}$ be a subset of $\mathcal{G}_1(V)$. For every $P \in \mathcal{X}$ we choose a non-zero vector $x_P \in P$ and consider the set $X := \{x_P\}_{P \in \mathcal{X}}$. It follows from Exercise 1.6 that $\mathcal{X}$ is an independent subset of Π_V if and only if X is an independent subset of V. Hence $\mathcal{X}$ is a base of the subspace $\mathcal{G}_1(S)$, $S \in \mathcal{G}(V)$ if and only if X is a base of the linear subspace S. Thus the dimension of $\mathcal{G}_1(S)$ is equal to $\dim S - 1$; in particular, Π_V is a projective space of dimension $\dim V - 1$.

Theorem 1.3. *If the dimension of a projective space is not less than 3 (the dimension can be infinite) then this projective space is isomorphic to the projective space associated with a certain vector space.*

Proof. See Chapter VII in [Baer (1952)]. □

Remark 1.2. For projective planes this fails. All projective spaces over division rings satisfy Desargues' axiom; but there exist non-Desarguesian projective planes.

Denote by $P^*(V)$ the Grassmannian consisting of all linear subspaces of codimension 1 in V (in other words, $P^*(V)$ coincides with $\mathcal{G}_{n-1}(V)$ or

$\mathcal{G}^1(V)$ if V is n-dimensional or infinite-dimensional, respectively). A subset $\mathcal{X} \subset P^*(V)$ is said to be a *line* if there is a linear subspace $S \subset V$ of codimension 2 such that $\mathcal{X}$ consists of all elements of $P^*(V)$ containing S. Let $\mathcal{L}^*(V)$ be the set of all such lines. An easy verification shows that

$$\Pi_V^* := (P^*(V), \mathcal{L}^*(V))$$

is a linear space; moreover, it is isomorphic to Π_{V^*} by the annihilator mapping (Corollary 1.1). The projective space Π_V^* is called *dual* to Π_V.

Exercise 1.7. Show that for every finite-dimensional subspace $\mathcal{S}$ of Π_V^* there exists $S \in \mathcal{G}^{\mathrm{fin}}(V)$ such that $\mathcal{S}$ consists of all elements of $P^*(V)$ containing S. *Hint:* use Corollary 1.1.

1.3   Semilinear mappings

Throughout the section we suppose that V and V' are left vector spaces over division rings R and R', respectively.

1.3.1   *Definitions*

We say that a mapping $l : V \to V'$ is *semilinear* if it is additive:

$$l(x + y) = l(x) + l(y) \quad \forall\, x, y \in V,$$

and there exists a homomorphism $\sigma : R \to R'$ such that

$$l(ax) = \sigma(a)l(x) \tag{1.1}$$

for all $a \in R$ and $x \in V$. In the case when $R = R'$ and σ is identity, we get a linear mapping. As for linear mappings, all vectors $x \in V$ satisfying $l(x) = 0$ form a linear subspace called the *kernel* of l and denoted by $\mathrm{Ker}\, l$. If l is non-zero ($\mathrm{Ker}\, l \neq V$) then there exists the unique homomorphism $\sigma : R \to R'$ satisfying (1.1) and the mapping l is said to be *σ-linear.*

Every semilinear mapping between real vector spaces is linear. In the general case, semilinear mappings form a more wide class.

A semilinear mapping of V to V' is called a *semilinear isomorphism* if it is bijective and the associated homomorphism of R to R' is an isomorphism. If $l : V \to V'$ is a σ-linear isomorphism then the inverse mapping l^{-1} is a σ^{-1}-linear isomorphism of V' to V. Semilinear isomorphisms map independent subsets to independent subsets and bases to bases. Semilinear isomorphisms of V to V' exist if and only if $\dim V = \dim V'$ and R is isomorphic to R'.

Exercise 1.8. Show that every semilinear bijection sending independent subsets to independent subsets is a semilinear isomorphism.

The following example shows that there exist semilinear bijections over non-surjective homomorphisms of division rings.

Example 1.3. The complexification mapping of $\mathbb{R}^{2n}$ to $\mathbb{C}^n$:
$$(x_1, y_1, \ldots, x_n, y_n) \to (x_1 + y_1 i, \ldots, x_n + y_n i)$$
is a semilinear bijection; the associated homomorphism is the natural embedding of $\mathbb{R}$ in $\mathbb{C}$.

The group of all semilinear automorphisms of V (semilinear isomorphisms of V to itself) is denoted by $\Gamma L(V)$. It contains $GL(V)$ as a subgroup. Consider the homomorphism of $\Gamma L(V)$ to $\mathrm{Aut}(R)$ which sends every σ-linear automorphism of V to σ. The kernel of this homomorphism is $GL(V)$. Hence $GL(V)$ is a normal subgroup of $\Gamma L(V)$ and the corresponding quotient group is isomorphic to $\mathrm{Aut}(R)$.

Example 1.4. The homothetic transformation $x \to ax$, $x \in V$, is a semilinear automorphism of V (the associated automorphism of R is $r \to ara^{-1}$). This transformation is linear only in the case when the scalar a belongs to the center of the division ring R. The group of all homothetic transformations $H(V)$ is isomorphic to the multiplicative group of the division ring; moreover, it is a normal subgroup of $\Gamma L(V)$.

1.3.2 *Mappings of Grassmannians induced by semilinear mappings*

Every semilinear mapping $l : V \to V'$ induces the mapping
$$(l)_1 : \mathcal{G}_1(V) \setminus \mathcal{G}_1(\mathrm{Ker}\, l) \to \mathcal{G}_1(V')$$
$$\langle x \rangle \to \langle l(x) \rangle, \quad x \notin \mathrm{Ker}\, l.$$
In the case when l is a semilinear isomorphism, this mapping is a collineation of Π_V to $\Pi_{V'}$.

If $l : V \to V'$ is a σ-linear mapping then for every non-zero scalar $a \in R'$ the mapping al is σ'-semilinear with
$$\sigma'(r) = a\sigma(r)a^{-1} \quad \forall\, r \in R,$$
and $(al)_1$ coincides with $(l)_1$.

Proposition 1.7. *Let $l : V \to V'$ be a semilinear mapping such that $l(V)$ contains two linearly independent vectors. If $s : V \to V'$ is a semilinear mapping satisfying $(l)_1 = (s)_1$ then $s = al$ for a certain scalar $a \in R'$.*

This statement is a direct consequence of the following lemma.

Lemma 1.2. *Let $l : V \to V'$ and $s : V \to V'$ be additive mappings such that*

$$s(x) \in \langle l(x) \rangle \quad \forall \, x \in V.$$

If $l(V)$ contains two linearly independent vectors then $s = al$ for a certain scalar $a \in R'$.

Proof. For every vector $x \in V$ there exists a scalar $a_x \in R'$ (possible zero) such that

$$s(x) = a_x l(x).$$

Suppose that vectors $l(x)$ and $l(y)$ both are non-zero. Then

$$a_x l(x) + a_y l(y) = s(x + y) = a_{x+y}(l(x) + l(y))$$

and $a_x = a_y = a_{x+y}$ if $l(x), l(y)$ are linearly independent. If $l(y)$ is a scalar multiple of $l(x)$ then we take any vector $z \in V$ such that $l(x)$ and $l(z)$ are linearly independent (by our hypothesis, this is possible) and get $a_x = a_z = a_y$. Therefore, for all vectors $x \in V$ satisfying $l(x) \neq 0$ the scalar a_x is a constant. The equality $l(x) = 0$ implies that $s(x) = 0$ and we have $s(x) = al(x)$ for every scalar $a \in R'$. $\qquad \square$

Proposition 1.8 ([Zick (1983)]). *Let $l : V \to V'$ be an additive mapping which satisfies*

$$l(\langle x \rangle) \subset \langle l(x) \rangle \quad \forall \, x \in V.$$

If $l(V)$ contains two linearly independent vectors then l is semilinear.

Proof. For a non-zero scalar $a \in R$ consider the mapping $x \to l(ax)$. Lemma 1.2 implies the existence of a scalar $\sigma(a) \in R'$ such that

$$l(ax) = \sigma(a)l(x) \quad \forall \, x \in V.$$

We set $\sigma(0) := 0$ and get a mapping $\sigma : R \to R'$. For any scalars $a, b \in R$ and any vector $x \in V$ we have

$$\sigma(a + b)l(x) = l(ax + bx) = l(ax) + l(bx) = (\sigma(a) + \sigma(b))l(x),$$

$$\sigma(ab)l(x) = l(abx) = \sigma(a)\sigma(b)l(x)$$

which means that σ is a homomorphism. $\qquad \square$

It was noted above that for every semilinear isomorphism $l : V \to V'$ the mapping $(l)_1$ is a collineation of Π_V to $\Pi_{V'}$. Conversely, we have the following.

Proposition 1.9. *Let* $l : V \to V'$ *be a semilinear mapping such that* $(l)_1$ *is a collineation of* Π_V *to* $\Pi_{V'}$. *Then* l *is a semilinear isomorphism of* V *to* V'.

Proof. Let $S \in \mathcal{G}_2(V)$ and $\{x, y\}$ be a base of S. Then $a \to \langle x + ay \rangle$ is a bijection of the division ring R to $\mathcal{G}_1(S) \setminus \{\langle y \rangle\}$. The mapping $(l)_1$ sends $\langle x + ay \rangle$ to

$$\langle l(x) + \sigma(a)l(y) \rangle,$$

where $\sigma : R \to R'$ is the homomorphism associated with l. Since the restriction of $(l)_1$ to the line $\mathcal{G}_1(S)$ is a bijection on the line $\mathcal{G}_1(l(S))$, the homomorphism σ is surjective; hence it is an isomorphism of R to R'. The latter implies that $l(V)$ is a linear subspace of V'; then $l(V) = V'$ (otherwise, $(l)_1$ is not surjective). Since $\operatorname{Ker} l = 0$, the mapping l is a semilinear isomorphism. $\qquad\square$

Every semilinear isomorphism of V to V' induces bijections between the Grassmannians of the same indices; in particular, it defines a collineation of Π_V^* to $\Pi_{V'}^*$.

Proposition 1.10. *Suppose that* $\mathcal{G}$ *and* $\mathcal{G}'$ *are Grassmannians of* V *and* V' *(respectively) with the same index. Let* $\bar{l} : \mathcal{G} \to \mathcal{G}'$ *and* $\bar{s} : \mathcal{G} \to \mathcal{G}'$ *be the bijections induced by semilinear isomorphisms* $l : V \to V'$ *and* $s : V \to V'$, *respectively. If* $\bar{l} = \bar{s}$ *then* $s = al$ *for a certain scalar* $a \in R'$.

Proof. Let $P \in \mathcal{G}_1(V)$. Consider the set $\mathcal{X}$ consisting of all elements of $\mathcal{G}$ containing P. Then

$$P = \bigcap_{S \in \mathcal{X}} S$$

and

$$l(P) = \bigcap_{S \in \bar{l}(\mathcal{X})} S, \quad s(P) = \bigcap_{S \in \bar{s}(\mathcal{X})} S.$$

Since $\bar{l} = \bar{s}$, we have $l(P) = s(P)$ for every $P \in \mathcal{G}_1(V)$. In other words, $(l)_1 = (s)_1$ and Proposition 1.7 gives the claim. $\qquad\square$

A semilinear injection $l : V \to V'$ is called a *semilinear embedding* of V in V' if it maps independent subsets to independent subsets. Semilinear embeddings of V in V' exist only in the case when $\dim V \leq \dim V'$ and R isomorphic to a certain division subring of R'.

Example 1.5. The mapping

$$(x_1, \ldots, x_n) \to (x_1 + 0i, \ldots, x_n + 0i)$$

is a semilinear embedding of $\mathbb{R}^n$ in $\mathbb{C}^n$. Similarly, the natural embedding of $\mathbb{Q}^n$ in $\mathbb{R}^n$ is a semilinear embedding.

Now we consider the mappings of Grassmannians induced by semilinear embeddings. We restrict ourselves to the case when $\dim V = n$ is finite (the infinite-dimensional case is similar).

Let $l : V \to V'$ be a semilinear embedding. Then for any linear subspace $S \subset V$ we have

$$\dim \langle l(S) \rangle = \dim S,$$

and the mapping

$$(l)_k : \mathcal{G}_k(V) \to \mathcal{G}_k(V')$$

$$S \to \langle l(S) \rangle$$

is defined for every $k \in \{1, \ldots, n-1\}$.

Exercise 1.9. Show that the following assertions are fulfilled:

(1) $(l)_k$ is injective for every $k \in \{1, \ldots, n-1\}$,
(2) $(l)_1$ is an embedding of Π_V in $\Pi_{V'}$,
(3) $(l)_{n-1}$ is an embedding of Π_V^* in $\Pi_{V'}^*$ if $\dim V' = n$,
(4) if $s : V \to V'$ is a semilinear embedding such that $(l)_k = (s)_k$ for a certain $k \in \{1, \ldots, n-1\}$ then s is a scalar multiple of l.

Proposition 1.11. *Let $l : V \to V'$ be a semilinear embedding such that $(l)_k$ is bijective for a certain $k \in \{1, \ldots, n-1\}$. Then l is a semilinear isomorphism.*

Proof. In the case when $k = 1$, our bijection maps any triple of collinear points of Π_V to collinear points of $\Pi_{V'}$ and any triple of non-collinear points to non-collinear points. Hence it is a collineation of Π_V to $\Pi_{V'}$. By Proposition 1.9, l is a semilinear isomorphism.

Suppose that $k > 1$. Let $S' \in \mathcal{G}_{k-1}(V')$. We choose $U_i' \in \mathcal{G}_k(V')$, $i = 1, 2$, such that

$$S' = U_1' \cap U_2'.$$

Since $(l)_k$ is bijective, there exist $U_i \in \mathcal{G}_k(V)$, $i = 1, 2$, satisfying

$$\langle l(U_i) \rangle = U_i'.$$

Then $U_1 \cap U_2$ belongs to $\mathcal{G}_{k-1}(V)$ (we leave the details for the reader) and

$$\langle l(U_1 \cap U_2) \rangle = S'.$$

Thus $(l)_{k-1}$ is surjective, and by the statement (1) of Exercise 1.9, it is bijective. Step by step, we establish that $(l)_1$ is bijective.   □

1.3.3   *Contragradient*

Now suppose that V and V' have the same finite dimension n. Recall that in this case $V = V^{**}$ and $V' = V'^{**}$. For every $x \in V$ and $x^* \in V^*$ we will write $x^* \cdot x$ instead of $x^*(x)$.

Let $u : V \to V'$ be a σ-linear isomorphism. The *adjoint* mapping

$$u^* : V'^* \to V^*$$

is defined by the condition

$$u^*(x^*) \cdot x = \sigma^{-1}(x^* \cdot u(x)) \quad \forall\, x \in V,\ x^* \in V'^*.$$

This is a (σ^{-1})-linear isomorphism and $u^{**} = u$. The inverse mapping

$$\check{u} := (u^*)^{-1} : V^* \to V'^*$$

is known as the *contragradient* of u. Since

$$(u^*)^{-1} = (u^{-1})^*,$$

the contragradient of the contragradient coincides with u. An easy verification shows that

$$\check{u}(x^*) \cdot u(x) = \sigma(x^* \cdot x) \quad \forall\, x \in V,\ x^* \in V^*.$$

Therefore, $\check{u}$ transfers the annihilator of a linear subspace $S \subset V$ to the annihilator of $u(S)$ and we get

$$u(U^0)^0 = \check{u}(U)$$

for every linear subspace $U \subset V^*$. So, the mapping

$$\mathcal{G}_k(V^*) \to \mathcal{G}_k(V'^*)$$

$$U \to u(U^0)^0$$

coincides with $(\check{u})_k$ for every $k \in \{1, \ldots, n-1\}$.

Exercise 1.10. Show that the contragradient mapping $u \to \check{u}$ is an isomorphism of $\Gamma\mathrm{L}(V)$ to $\Gamma\mathrm{L}(V^*)$ transferring $\mathrm{GL}(V)$ to $\mathrm{GL}(V^*)$.

1.4   Fundamental Theorem of Projective Geometry

1.4.1   *Main theorem and corollaries*

Let V and V' be left vector spaces over division rings R and R', respectively. The dimensions of V and V' are assumed to be not less than 3. Let also $l : V \to V'$ be a semilinear injection. The mapping

$$(l)_1 : \mathcal{G}_1(V) \to \mathcal{G}_1(V')$$

$$P \to \langle l(P) \rangle$$

does not need to be injective, see Example 1.3. For any $P, P_1, P_2 \in \mathcal{G}_1(V)$ satisfying $P \subset P_1 + P_2$ (a triple of collinear points of Π_V) we have

$$\langle l(P) \rangle \subset \langle l(P_1) \rangle + \langle l(P_2) \rangle.$$

If $\langle l(P_1) \rangle = \langle l(P_2) \rangle$ then the line $\mathcal{G}_1(P_1 + P_2)$ goes to a point. In the case when $\langle l(P_1) \rangle$ and $\langle l(P_2) \rangle$ are distinct, the restriction of $(l)_1$ to $\mathcal{G}_1(P_1 + P_2)$ is a bijection on a subset of the line $\mathcal{G}_1(\langle l(P_1 + P_2) \rangle)$.

The following generalized version of the Fundamental Theorem of Projective Geometry was proved in [Faure and Frolicher (1994)] and [Havlicek (1994)], independently.

Theorem 1.4. *Suppose that a mapping $f : \mathcal{G}_1(V) \to \mathcal{G}_1(V')$ satisfies the following conditions: for any $P, P_1, P_2 \in \mathcal{G}_1(V)$*

$$P \subset P_1 + P_2 \implies f(P) \subset f(P_1) + f(P_2)$$

and $f(\mathcal{G}_1(V))$ is not contained in a line. Then f is induced by a semilinear injection of V to V'.

Theorem 1.4 together with Proposition 1.9 give the classical version of the Fundamental Theorem of Projective Geometry.

Corollary 1.2. *Every collineation of Π_V to $\Pi_{V'}$ is induced by a semilinear isomorphism of V to V'.*

Remark 1.3. Denote by $P\Gamma L(V)$ the group formed by all collineations of the projective space Π_V to itself. Consider the homomorphism of $\Gamma L(V)$ to $P\Gamma L(V)$ which sends l to $(l)_1$. The Fundamental Theorem of Projective Geometry (Corollary 1.2) states that this homomorphism is surjective. By Proposition 1.7, the kernel of the homomorphism is $H(V)$ (Example 1.4). Therefore, $P\Gamma L(V)$ is isomorphic to the quotient group $\Gamma L(V)/H(V)$. In the case when V is finite-dimensional, the contragradient isomorphism of $\Gamma L(V)$ to $\Gamma L(V^*)$ (Exercise 1.10) sends $H(V)$ to $H(V^*)$; hence it induces an isomorphism between $P\Gamma L(V)$ and $P\Gamma L(V^*)$.

Corollary 1.3. *If V and V' have the same finite dimension then every semicollineation of Π_V to $\Pi_{V'}$ is a collineation.*

Proof. Every semicollineation of Π_V to $\Pi_{V'}$ satisfies the conditions of Theorem 1.4; hence it is induced by a semilinear mapping $l : V \to V'$. Our semicollineation maps every triple of collinear points to collinear points and it is sufficient to show that triples of non-collinear points go to non-collinear points.

If this fails then there exist linearly independent vectors $x, y, z \in V$ such that $l(x), l(y), l(z)$ are linearly dependent. Let B be a base of V containing x, y, z. Since $\dim V = \dim V'$ is finite, $l(B)$ spans a proper subspace of V' which contradicts $\langle l(B) \rangle = \langle l(V) \rangle = V'$. $\qquad\square$

Corollary 1.3 will be generalized in Section 3.5.

Problem 1.1. Is there a semicollineation of Π_V to $\Pi_{V'}$ in the case when $\dim V > \dim V'$?

It was noted in the previous section that semilinear isomorphisms of V to V' induce collineations of Π_V^* to $\Pi_{V'}^*$. We can show that there exist no other collineations of Π_V^* to $\Pi_{V'}^*$ only in the case when our vector spaces are finite-dimensional.

Corollary 1.4. *If V and V' are finite-dimensional then every collineation of Π_V^* to $\Pi_{V'}^*$ is induced by a semilinear isomorphism of V to V'.*

Proof. Let f be a collineation of Π_V^* to $\Pi_{V'}^*$. It can be considered as a collineation of Π_{V^*} to $\Pi_{V'^*}$. By the Fundamental Theorem of Projective Geometry, the latter collineation is induced by a semilinear isomorphism $u : V^* \to V'^*$ and

$$f(S) = u(S^0)^0$$

for every linear subspace $S \subset V$ of codimension 1. By Subsection 1.3.3, we have $f = (\check{u})_1$ in the finite-dimension case. $\qquad\square$

Exercise 1.11. Show that every semicollineation of Π_V^* to $\Pi_{V'}^*$ is a collineation if $\dim V = \dim V' < \infty$.

Let X be a partially ordered set. The *supremum* of a subset $Y \subset X$ is the least element of X which is greater than or equal to all elements of Y. Similarly, the *infimum* of Y is the greatest element of X which is less than or equal to every element of Y. The supremum and infimum of Y

are denoted by $\sup Y$ and $\inf Y$, respectively. If $\sup Y$ ($\inf Y$) exists then it is unique. The set X is called a *lattice* if every subset consisting of two elements has the supremum and infimum. The set $\mathcal{G}(V)$ is partially ordered by the inclusion relation. Since for all $S, U \in \mathcal{G}(V)$ we have

$$\sup\{S, U\} = S + U \quad \text{and} \quad \inf\{S, U\} = S \cap U,$$

it is a lattice. Every semilinear isomorphism of V to V' induces an order preserving bijection of $\mathcal{G}(V)$ to $\mathcal{G}(V')$ (an isomorphism between the lattices). The classical Fundamental Theorem of Projective Geometry can be reformulated in the following form (see, for example, [Baer (1952)]).

Corollary 1.5. *Let $f : \mathcal{G}(V) \to \mathcal{G}(V')$ be an order preserving bijection: for all $S, U \in \mathcal{G}(V)$*

$$S \subset U \iff f(S) \subset f(U).$$

Then f is induced by a semilinear isomorphism of V to V'.

Proof. It is not difficult to prove that f transfers $\mathcal{G}_1(V)$ and $\mathcal{G}_2(V)$ to $\mathcal{G}_1(V')$ and $\mathcal{G}_2(V')$, respectively. Hence the restriction of f to $\mathcal{G}_1(V)$ is a collineation of Π_V to $\Pi_{V'}$ and there exists a semilinear isomorphism $l : V \to V'$ such that $f(P) = l(P)$ for all $P \in \mathcal{G}_1(V)$. Then for every $S \in \mathcal{G}(V)$

$$f(\mathcal{G}_1(S)) = \mathcal{G}_1(f(S))$$

coincides with

$$l(\mathcal{G}_1(S)) = \mathcal{G}_1(l(S))$$

and we get $f(S) = l(S)$. $\qquad\qquad\square$

1.4.2 *Proof of Theorem 1.4*

Let $f : \mathcal{G}_1(V) \to \mathcal{G}_1(V')$ be a mapping which satisfies the conditions of Theorem 1.4: for any $P, P_1, P_2 \in \mathcal{G}_1(V)$

$$P \subset P_1 + P_2 \implies f(P) \subset f(P_1) + f(P_2)$$

and $f(\mathcal{G}_1(V))$ is not contained in a line. This means that the restriction of f to every line is a constant or a bijection to a subset of a line.

Lemma 1.3. *Let $v_1, v_2, v_3 \in V$ and $w_1, w_2, w_3 \in V'$ be vectors satisfying the following conditions:*

- $f(\langle v_i \rangle) = \langle w_i \rangle$ *for $i = 1, 2, 3$,*

- $f(\langle v_1 + v_2 \rangle) = \langle w_1 + w_2 \rangle$ and $f(\langle v_1 + v_3 \rangle) = \langle w_1 + w_3 \rangle$,
- w_1, w_2, w_3 are linearly independent.

Then

$$f(\langle v_2 + v_3 \rangle) = \langle w_2 + w_3 \rangle \quad and \quad f(\langle v_1 + v_2 + v_3 \rangle) = \langle w_1 + w_2 + w_3 \rangle.$$

Proof. It is clear that v_1, v_2, v_3 are linearly independent and $\langle v_1 + v_2 + v_3 \rangle$ is the intersection of

$$\langle v_1 + v_2 \rangle + \langle v_3 \rangle \quad and \quad \langle v_1 + v_3 \rangle + \langle v_2 \rangle.$$

Then $f(\langle v_1 + v_2 + v_3 \rangle)$ is contained in the intersection of

$$\langle w_1 + w_2 \rangle + \langle w_3 \rangle \quad and \quad \langle w_1 + w_3 \rangle + \langle w_2 \rangle.$$

The latter intersection coincides with $\langle w_1 + w_2 + w_3 \rangle$ and we get the second equality. Similarly,

$$\langle v_2 + v_3 \rangle = (\langle v_2 \rangle + \langle v_3 \rangle) \cap (\langle v_1 \rangle + \langle v_1 + v_2 + v_3 \rangle)$$

and $f(\langle v_2 + v_3 \rangle)$ is contained

$$(\langle w_2 \rangle + \langle w_3 \rangle) \cap (\langle w_1 \rangle + \langle w_1 + w_2 + w_3 \rangle)$$

which gives the first equality. $\qquad\square$

Lemma 1.4. *Let $P, Q \in \mathcal{G}_1(V)$ and $x \in P + Q$ with $x \notin Q$. Then there exists a unique vector $y \in Q$ such that $P = \langle x + y \rangle$.*

Proof. Trivial. $\qquad\square$

By our hypothesis, there exist vectors $x_1, x_2, x_3 \in V$ such that

$$f(\langle x_1 \rangle), f(\langle x_2 \rangle), f(\langle x_3 \rangle)$$

are not contained in a line. Our first step is to establish the existence of vectors $y_1, y_2, y_3 \in V'$ satisfying the conditions

$$f(\langle x_i \rangle) = \langle y_i \rangle \quad and \quad f(\langle x_i + x_j \rangle) = \langle y_i + y_j \rangle.$$

Proof. Let us take any non-zero vector $y_1 \in f(\langle x_1 \rangle)$. Since $\langle x_1 \rangle$ is contained in $\langle x_1 + x_2 \rangle + \langle x_2 \rangle$,

$$y_1 \in f(\langle x_1 \rangle) \subset f(\langle x_1 + x_2 \rangle) + f(\langle x_2 \rangle).$$

Lemma 1.4 implies the existence of $y_2 \in f(\langle x_2 \rangle)$ such that

$$f(\langle x_1 + x_2 \rangle) = \langle y_1 + y_2 \rangle.$$

If y_2 is zero then $f(\langle x_1 + x_2 \rangle)$ coincides with $f(\langle x_1 \rangle)$ and the restriction of f to the line associated with the linear subspace

$$\langle x_1 + x_2 \rangle + \langle x_1 \rangle = \langle x_1 \rangle + \langle x_2 \rangle$$

is constant; in this case, we get $f(\langle x_1 \rangle) = f(\langle x_2 \rangle)$ which is impossible. Thus $y_2 \neq 0$. Similarly, we establish the existence of a non-zero vector $y_3 \in f(\langle x_3 \rangle)$ satisfying

$$f(\langle x_1 + x_3 \rangle) = \langle y_1 + y_3 \rangle.$$

By Lemma 1.3, the vectors y_1, y_2, y_3 are as required. $\square$

Let $x \in V \setminus \{0\}$. We choose $i \in \{1, 2, 3\}$ such that $f(\langle x \rangle) \neq f(\langle x_i \rangle)$. The restriction of f to the line associated with the linear subspace

$$\langle x \rangle + \langle x_i \rangle = \langle x + x_i \rangle + \langle x \rangle$$

is injective and we apply Lemma 1.4 to

$$y_i \in f(\langle x_i \rangle) \subset f(\langle x_i + x \rangle) + f(\langle x \rangle).$$

So, there exists a vector $l(x) \in f(\langle x \rangle)$ such that

$$f(\langle x_i + x \rangle) = \langle y_i + l(x) \rangle.$$

Since $f(\langle x_i + x \rangle) \neq f(\langle x_i \rangle)$, we have $l(x) \neq 0$. We set $l(0) := 0$ and assert that $l : V \to V'$ is a semilinear mapping.

Show that the mapping $l : V \to V'$ is well-defined (the definition of the vector $l(x)$ does not depend on the choice of the number $i \in \{1, 2, 3\}$).

Proof. Suppose that

$$f(\langle x_1 \rangle) \neq f(\langle x \rangle) \neq f(\langle x_2 \rangle).$$

There exists a non-zero vector $l(x) \in f(\langle x \rangle)$ satisfying

$$f(\langle x_1 + x \rangle) = \langle y_1 + l(x) \rangle.$$

We need to show that

$$f(\langle x_2 + x \rangle) = \langle y_2 + l(x) \rangle.$$

If $f(\langle x \rangle)$ is not contained in $f(\langle x_1 \rangle) + f(\langle x_2 \rangle)$ then the required equality follows from Lemma 1.3. In the case when

$$f(\langle x \rangle) \subset f(\langle x_1 \rangle) + f(\langle x_2 \rangle),$$

we consider successively the triples x_1, x_3, x and x_3, x_2, x. $\square$

Our next step is the equality $l(v + w) = l(v) + l(w)$ for all $v, w \in V$ (the mapping l is additive).

Proof. First consider the case when $l(v)$ and $l(w)$ are linearly independent. We choose $i \in \{1, 2, 3\}$ such that $f(\langle x_i \rangle)$ is not contained in $f(\langle v \rangle) + f(\langle w \rangle)$ (this is possible, since $f(\langle x_1 \rangle), f(\langle x_2 \rangle), f(\langle x_3 \rangle)$ are not contained in a line). We have

$$f(\langle x_i + v \rangle) = \langle y_i + l(v) \rangle \quad \text{and} \quad f(\langle x_i + w \rangle) = \langle y_i + l(w) \rangle,$$

and Lemma 1.3 implies that

$$f(\langle v + w \rangle) = \langle l(v) + l(w) \rangle \quad \text{and} \quad f(\langle x_i + v + w \rangle) = \langle y_i + l(v) + l(w) \rangle.$$

Thus

$$\langle l(v + w) \rangle = \langle l(v) + l(w) \rangle \quad \text{and} \quad \langle y_i + l(v + w) \rangle = \langle y_i + l(v) + l(w) \rangle.$$

By the first equality, $l(v + w) = a(l(v) + l(w))$ for a certain scalar $a \in R'$; the second equality guarantees that $a = 1$.

Now suppose that $l(w)$ is a scalar multiple of $l(v)$. Then

$$f(\langle v \rangle) = f(\langle w \rangle) = f(\langle v + w \rangle).$$

We take any vector $z \in V$ such that $l(w), l(z)$ are linearly independent. Then $l(w + z) = l(w) + l(z)$; moreover, $l(v + w), l(z)$ are linearly independent and

$$l(v + w + z) = l(v + w) + l(z). \tag{1.2}$$

Since $f(\langle w \rangle) \neq f(\langle z \rangle)$, the restriction of f to the line associated with the linear subspace $\langle w \rangle + \langle z \rangle$ is injective and

$$f(\langle v \rangle) = f(\langle w \rangle) \neq f(\langle w + z \rangle).$$

This means that $l(v), l(w + z)$ are linearly independent; hence

$$l(v + w + z) = l(v) + l(w + z) = l(v) + l(w) + l(z)$$

and (1.2) gives the claim. $\square$

For every non-zero vector $x \in V$ and any non-zero scalar $a \in R$ we have

$$\langle l(ax) \rangle = f(\langle ax \rangle) = f(\langle x \rangle) = \langle l(x) \rangle.$$

In other words, $l(\langle x \rangle) \subset \langle l(x) \rangle$ and the mapping $l : V \to V'$ satisfies the conditions of Proposition 1.8. Therefore, l is semilinear and $f = (l)_1$.

Remark 1.4. This proof was taken from the paper [Faure (2002)] dedicated to A. Frölicher.

1.4.3   *Fundamental Theorem for normed spaces*

Let F be the field of real or complex numbers. Let also N be a *normed space* over F; this means that N is a vector space (over F) with a real-valued function $x \to ||x||$ called a *norm* and satisfying the following conditions:

- $||x|| \geq 0$ for all $x \in N$, and $||x|| = 0$ implies that $x = 0$,
- $||x + y|| \leq ||x|| + ||y||$ for all $x, y \in N$,
- $||ax|| = |a| \cdot ||x||$ for every vector $x \in N$ and every scalar $a \in F$.

Our normed space is, in a natural way, a matric space (hence it is also a topological space); the metric is defined by

$$d(x, y) := ||x - y|| \quad \forall \, x, y \in N.$$

Denote by $\mathcal{G}_{\mathrm{cl}}(N)$ the set of all closed linear subspaces of N. Every finite-dimensional linear subspace is closed; hence $\mathcal{G}_{\mathrm{cl}}(N)$ coincides with $\mathcal{G}(N)$ if the dimension of N is finite.

The sum of two closed linear subspaces does not need to be closed. For linear subspaces $S, U \subset N$ we denote by $S \dot{+} U$ the closure of $S + U$; since the closure of a linear subspace is a linear subspace, we have $S \dot{+} U \in \mathcal{G}_{\mathrm{cl}}(N)$. The set $\mathcal{G}_{\mathrm{cl}}(N)$ is partially ordered by the inclusion relation; moreover, for all $S, U \in \mathcal{G}_{\mathrm{cl}}(V)$

$$\sup\{S, U\} = S \dot{+} U \quad \text{and} \quad \inf\{S, U\} = S \cap U.$$

So, $\mathcal{G}_{\mathrm{cl}}(N)$ is a lattice. Note that $\mathcal{G}_{\mathrm{cl}}(N)$ is not a sublattice of the lattice $\mathcal{G}(N)$ if N is infinite-dimensional.

Example 1.6. The vector space F^n is normed by

$$||x|| := (|x_1|^2 + \cdots + |x_n|^2)^{1/2}, \quad x = (x_1, \ldots, x_n);$$

in particular, $(F, |\cdot|)$ is a 1-dimensional normed space.

Example 1.7. The vector space $C_F([0, 1])$ consisting of all continuous functions $g : [0, 1] \to F$ is normed by

$$||g|| := \sup\{ |g(x)| \ : \ x \in [0, 1] \}.$$

Example 1.8. Let $p \in \mathbb{N}$ and $l_p(F)$ be the vector space formed by all sequences $x = \{x_n\}_{n \in \mathbb{N}} \subset F$ satisfying

$$||x||_p := \left(\sum_{n=1}^{\infty} |x_n|^p \right)^{1/p} < \infty.$$

Then $(l_p(F), ||\cdot||_p)$ is a normed space.

Let M be other normed vector space over F. We will investigate order preserving bijections of $\mathcal{G}_{\mathrm{cl}}(N)$ to $\mathcal{G}_{\mathrm{cl}}(M)$ (isomorphisms between the lattices). Recall that a bijection h between two topological spaces is called a *homeomorphism* if h and h^{-1} both are continuous. Clearly, every linear homeomorphism of N to M induces an order preserving bijection of $\mathcal{G}_{\mathrm{cl}}(N)$ to $\mathcal{G}_{\mathrm{cl}}(M)$.

Now suppose that our normed spaces are complex. If a semilinear mapping $l : N \to M$ is continuous then the associated homomorphism $\sigma : \mathbb{C} \to \mathbb{C}$ is continuous. By Exercise 1.3, σ is identity or the complex conjugate mapping. In what follows a semilinear mapping between complex vector spaces will be called *conjugate-linear* if the associated homomorphism of $\mathbb{C}$ to itself is the complex conjugate mapping. Every conjugate-linear homeomorphism of N to M induces an order preserving bijection of $\mathcal{G}_{\mathrm{cl}}(N)$ to $\mathcal{G}_{\mathrm{cl}}(M)$.

Theorem 1.5 (G. W. Mackey, P. A. Fillmore, W. E. Longstaff).
Let $f : \mathcal{G}_{\mathrm{cl}}(N) \to \mathcal{G}_{\mathrm{cl}}(M)$ be an order preserving bijection:

$$S \subset U \iff f(S) \subset f(U)$$

for all $S, U \in \mathcal{G}_{\mathrm{cl}}(N)$. If N and M are real normed spaces then f is induced by a linear homeomorphism of N to M. In the case when N and M are infinite-dimensional complex normed spaces, f is induced by a linear or conjugate-linear homeomorphism of N to M.

Remark 1.5. If N and M are finite-dimensional complex normed spaces then every semilinear isomorphism of N to M induces an order preserving bijection of $\mathcal{G}(N) = \mathcal{G}_{\mathrm{cl}}(N)$ to $\mathcal{G}(M) = \mathcal{G}_{\mathrm{cl}}(M)$.

Theorem 1.5 was first proved in [Mackey (1942)] for real normed spaces. The complex version of this result was given in [Fillmore and Longstaff (1984)].

1.4.4 *Proof of Theorem 1.5*

We need some elementary facts concerning linear and conjugate-linear mappings. Let N and M be normed spaces over $F = \mathbb{R}, \mathbb{C}$. For a linear mapping $l : N \to M$ (if $M = F$ then l is a linear functional) the following conditions are equivalent:

- l is continuous,

- l transfers bounded subsets to bounded subsets (a subset X is bounded if there exists a real non-negative number a such that $||x|| \leq a$ for all $x \in X$).

In the complex case, the same holds for conjugate-linear mappings. In [Rudin (1973)] (Section 1.32) this statement is proved for linear mappings; for conjugate-linear mappings the proof is similar. If $l : N \to M$ is a continuous linear or conjugate-linear mapping then the number

$$||l|| := \sup\{ ||l(x)|| : x \in N, ||x|| \leq 1 \}$$

is finite and called the *norm* of l; it is clear that

$$||l(x)|| \leq ||l|| \cdot ||x|| \quad \text{and} \quad ||al|| = |a| \cdot ||l||$$

for all $x \in N$ and $a \in F$.

Now we start to prove the theorem. Let $f : \mathcal{G}_{cl}(N) \to \mathcal{G}_{cl}(M)$ be an order preserving bijection. Then f transfers $\mathcal{G}_1(N)$ and $\mathcal{G}_2(N)$ to $\mathcal{G}_1(M)$ and $\mathcal{G}_2(M)$, respectively. This means that the restriction of f to $\mathcal{G}_1(N)$ is a collineation of Π_N to Π_M. By the Fundamental Theorem of Projective Geometry, it is induced by a certain semilinear isomorphism $l : N \to M$. As in the proof of Corollary 1.5, we establish that

$$f(S) = l(S) \quad \forall\, S \in \mathcal{G}_{cl}(N).$$

We need to show that l is a homeomorphism of N to M.

Real case. The mapping l is linear. Let $v : M \to \mathbb{R}$ be a non-zero continuous linear functional. Then $\operatorname{Ker} v$ is a closed linear subspace of codimension 1 and

$$S := l^{-1}(\operatorname{Ker} v)$$

is a closed linear subspace of codimension 1 in N. For every closed linear subspace of codimension 1 there exists a continuous linear functional whose kernel coincides with this linear subspace (a consequence of the Hahn-Banach Theorem, see Section 3.5 in [Rudin (1973)]). Consider a continuous linear functional $w : N \to \mathbb{R}$ such that $\operatorname{Ker} w = S$. We fix $z \in N$ satisfying $w(z) = 1$. Every $x \in N$ can be presented in the form $x = y + w(x)z$, where $y \in S$. Then

$$v(l(x)) = v(l(y)) + v(w(x)l(z)) = w(x)v(l(z)) \tag{1.3}$$

(since $l(y) \in \operatorname{Ker} v$).

Let X be a bounded subset of N. Then $w(X)$ is a bounded subset of $\mathbb{R}$ and (1.3) guarantees that the same holds for $v(l(X))$. Since v is arbitrary taken, the set $v(l(X))$ is bounded for every continuous linear functional $v : M \to \mathbb{R}$, in other words, $l(X)$ is weakly bounded. In a normed space every weakly bounded subset is bounded (Section 3.18 in [Rudin (1973)]). Thus l transfers bounded subsets to bounded subsets; hence it is continuous. Similarly, we establish that l^{-1} is continuous.

Complex case. The normed spaces are assumed to be infinite-dimensional. Let σ be the automorphism of $\mathbb{C}$ associated with l. In this case, we have

$$v(l(x)) = \sigma(w(x))v(l(z))$$

instead of (1.3). The arguments given above work only in the case when σ is identity or the complex conjugate mapping. Therefore, we need to show that l is linear or conjugate-linear. This fact is a direct consequence of Kakutani–Mackey's result which will be proved below (Proposition 1.12).

Lemma 1.5. *Let σ be an automorphism of $\mathbb{C}$. If for every sequence of complex numbers $z_n \to 0$ the sequence $\{\sigma(z_n)\}_{n \in \mathbb{N}}$ is bounded then σ is continuous.*

Proof. By the additivity of σ, it is sufficient to verify that σ is continuous in 0. Indeed, if $z_n \to z$ then $(z_n - z) \to 0$ and

$$\sigma(z_n - z) \to 0 \implies \sigma(z_n) \to \sigma(z).$$

Suppose that σ is non-continuous in 0. Then there exist a sequence $z_n \to 0$ and a real number $a > 0$ satisfying

$$|\sigma(z_n)| > a \quad \forall\, n \in \mathbb{N}.$$

Clearly, $\{z_n\}_{n \in \mathbb{N}}$ contains a subsequence $\{z'_n\}_{n \in \mathbb{N}}$ such that $|z'_n| < 1/n^2$ for every $n \in \mathbb{N}$, in other words, $nz'_n \to 0$. Since the restriction of σ to $\mathbb{N}$ is identity,

$$|\sigma(nz'_n)| = n|\sigma(z'_n)| > na \quad \forall\, n \in \mathbb{N}$$

and the sequence $\{\sigma(nz'_n)\}_{n \in \mathbb{N}}$ is unbounded, a contradiction. $\square$

Lemma 1.6. *If N is infinite-dimensional then there exist an independent subset $\{x_n\}_{n \in \mathbb{N}} \subset N$ and a sequence of continuous linear functionals $\{v_n\}_{n \in \mathbb{N}}$ of N such that*

$$v_i(x_j) = \delta_{ij}. \tag{1.4}$$

Proof. It is not difficult to choose two linearly independent vectors $x_1, x_2 \in N$ and two continuous linear functionals v_1, v_2 of N satisfying (1.4). Suppose that (1.4) holds for linearly independent vectors $x_1, \ldots, x_n$ and continuous linear functionals $v_1, \ldots, v_n$. Then there exists a continuous linear functional v_{n+1} such that

- all x_i belong to $\mathrm{Ker}\, v_{n+1}$,
- $v_{n+1}(x'_{n+1}) = 1$ for a certain vector x'_{n+1} which does not belong to $\langle x_1, \ldots, x_n \rangle$

(see Section 3.5 in [Rudin (1973)]). We define

$$x_{n+1} := x'_{n+1} - \sum_{i=1}^{n} v_i(x'_{n+1})x_i.$$

Then $v_{n+1}(x_{n+1}) = 1$ and $v_i(x_{n+1}) = 0$ for all $i \le n$. $\qquad\square$

Lemma 1.7. *If N is infinite-dimensional then it contains an independent subset $\{x_n\}_{n \in \mathbb{N}}$ satisfying the following condition: for every bounded sequence of scalars $\{a_n\}_{n \in \mathbb{N}}$ there exists a continuous linear functional $v : N \to F$ such that*

$$v(x_n) = a_n \quad \forall\, n \in \mathbb{N}.$$

Proof. Let $X = \{x_n\}_{n \in \mathbb{N}}$ and $\{v_n\}_{n \in \mathbb{N}}$ be as in the previous lemma. We can assume that

$$||v_n|| = 1/2^n \quad \forall\, n \in \mathbb{N}$$

(indeed, for every $n \in \mathbb{N}$ there exists a scalar b_n such that $||b_n v_n|| = 1/2^n$ and we can take $x'_n \in \langle x_n \rangle$ satisfying $b_n v_n(x'_n) = 1$).

Let $\{a_n\}_{n \in \mathbb{N}}$ be a bounded sequence of scalars and $a = \sup |a_i|$. For every vector

$$x = v_1(x)x_1 + \cdots + v_n(x)x_n \in \langle X \rangle$$

we define

$$v(x) := v_1(x)a_1 + \cdots + v_n(x)a_n.$$

Then

$$|v(x)| \le |a_1| \cdot ||v_1|| \cdot ||x|| + \cdots + |a_n| \cdot ||v_n|| \cdot ||x||$$

$$\le a||x||(1/2 + \cdots + 1/2^n) < a||x||.$$

So, $v : \langle X \rangle \to F$ is a continuous linear functional; by the Hahn-Banach Theorem, it can be extended to a continuous linear functional of N. $\qquad\square$

Proposition 1.12 ([Kakutani and Mackey (1946)]). *Suppose that N and M are infinite-dimensional complex normed spaces. Let $l : N \to M$ be a semilinear isomorphism which sends closed linear subspaces of codimension 1 to closed linear subspaces. Then l is linear or conjugate-linear.*

Proof. Let σ be the automorphism of $\mathbb{C}$ associated with l. Let also $\{x_n\}_{n\in\mathbb{N}}$ be a subset of N with the property described in the previous lemma. Show that for every sequence of complex numbers $a_n \to 0$ the sequence $\{\sigma(a_n)\}_{n\in\mathbb{N}}$ is bounded.

If $\{\sigma(a_n)\}_{n\in\mathbb{N}}$ is unbounded then $\{a_n\}_{n\in\mathbb{N}}$ contains a subsequence $\{b_n\}_{n\in\mathbb{N}}$ such that

$$|\sigma(b_n)| \geq n \|l(x_n)\| \quad \forall\, n \in \mathbb{N}. \tag{1.5}$$

Let $v : N \to \mathbb{C}$ be a continuous linear functional satisfying $v(x_n) = b_n$ for every n. We take $z \in N$ such that $v(z) = 1$. Then $x_n = y_n + b_n z$, where $y_n \in \operatorname{Ker} v$. We have

$$l(x_n)/\sigma(b_n) = l(y_n/b_n) + l(z).$$

By (1.5),

$$l(x_n)/\sigma(b_n) \to 0 \ \text{ and } \ l(-y_n/b_n) \to l(z).$$

Hence $l(z)$ belongs to the closure of $l(\operatorname{Ker} v)$. $\operatorname{Ker} v$ is a closed linear subspace of codimension 1 and the linear subspace $l(\operatorname{Ker} v)$ is, by our hypothesis, closed. Thus $l(z)$ belongs to $l(\operatorname{Ker} v)$. Since l is bijective, we get $z \in \operatorname{Ker} v$ which contradicts $v(z) = 1$.

By Lemma 1.5, σ is continuous. This means that it is identity or the complex conjugate mapping. $\qquad\square$

1.5 Reflexive forms and polarities

1.5.1 *Sesquilinear forms*

Let V be a left vector space over a division ring R and $\sigma : R \to R$ be an anti-automorphism (an isomorphism of R to the opposite division ring R^*). We say that

$$\Omega : V \times V \to R$$

is a *sesquilinear form* over σ or simple a *σ-form* if

$$\Omega(x + y, z) = \Omega(x, z) + \Omega(y, z), \ \ \Omega(z, x + y) = \Omega(z, x) + \Omega(z, y),$$

$$\Omega(ax, by) = a\Omega(x, y)\sigma(b)$$

for all vectors $x, y, z \in V$ and all scalars $a, b \in R$; if σ is identity (this is possible only in the commutative case) then Ω is a usual *bilinear form*.

For every $y \in V$ the mapping $x \to \Omega(x, y)$ is a linear functional of V and we get a σ-linear mapping of V to V^*. Conversely, every semilinear mapping $l : V \to V^*$ gives a sesquilinear form

$$(x, y) \to l(y) \cdot x.$$

Therefore, there is a one-to-one correspondence between sesquilinear forms on V and semilinear mappings of V to V^*.

Our form Ω is said to be *non-degenerate* if the associated semilinear mapping of V to V^* is injective (if V is finite-dimensional then $\dim V = \dim V^*$ and it is a semilinear isomorphism). This condition is equivalent to the fact that for every non-zero $y \in V$ there exists $x \in V$ such that $\Omega(x, y) \neq 0$.

1.5.2 Reflexive forms

Suppose that $\dim V = n$ is finite. A sesquilinear form $\Omega : V \times V \to R$ is called *reflexive* if

$$\Omega(x, y) = 0 \implies \Omega(y, x) = 0$$

for every $x, y \in V$. We give a few examples.

Example 1.9. A non-zero sesquilinear form $\Omega : V \times V \to R$ is said to be *symmetric* or *skew-symmetric* if

$$\Omega(x, y) = \Omega(y, x) \quad \forall\, x, y \in V$$

or

$$\Omega(x, y) = -\Omega(y, x) \quad \forall\, x, y \in V,$$

respectively. In each of these cases, the associated anti-automorphism is identity (we leave the verification to the reader); hence R is commutative. A sesquilinear form $\Theta : V \times V \to R$ is called *alternating* if

$$\Theta(x, x) = 0 \quad \forall\, x \in V.$$

Every alternating form is skew-symmetric. Conversely, if the characteristic of R is not equal to 2 then every skew-symmetric form on V is alternating. In the case of characteristic 2, the classes of symmetric and skew-symmetric forms are coincident. Non-degenerate alternating forms exist only on even-dimensional vector spaces over fields.

Example 1.10. Let σ be a non-identity anti-automorphism of R. We say that a non-zero sesquilinear form $\Omega : V \times V \to R$ is σ-*Hermitian* or *skew* σ-*Hermitian* if

$$\Omega(x, y) = \sigma(\Omega(y, x)) \quad \forall\, x, y \in V$$

or

$$\Omega(x, y) = -\sigma(\Omega(y, x)) \quad \forall\, x, y \in V,$$

respectively. An easy verification shows that Ω is a σ-form and $\sigma^2 = 1_R$ in each of these cases.

It is clear that the forms considered above are reflexive. Also note that for every reflexive σ-form $\Omega : V \times V \to R$ and every non-zero scalar $a \in R$ the scalar multiple

$$(x, y) \to \Omega(x, y)a$$

is a reflexive σ'-form with

$$\sigma'(b) = a^{-1}\sigma(b)a \quad \forall\, b \in R.$$

There is a complete description of all non-degenerate reflexive forms.

Theorem 1.6 (G. Birkhoff, J. von Neumann). *If $\Omega : V \times V \to R$ is a non-degenerate reflexive form then one of the following possibilities is realized:*

- *R is commutative and Ω is symmetric or alternating,*
- *Ω is a scalar multiple of a Hermitian form.*

Proof. See, for example, Section 1.6 in [Dieudonné (1971)] or Chapter 7 in [Taylor (1992)]. $\qquad\square$

Let $\Omega : V \times V \to R$ be a reflexive form. We say that two vectors $x, y \in V$ are *orthogonal* and write $x \perp y$ if $\Omega(x, y) = 0$. The orthogonality relation is symmetric (by the reflexivity). If X and Y are subsets of V then $X \perp Y$ means that every $x \in X$ is orthogonal to all $y \in Y$; in this case, we say that the subsets X and Y are *orthogonal*. The linear subspace consisting of all vectors orthogonal to a subset $X \subset V$ is called the *orthogonal complement* of X and denoted by $X^\perp$.

A non-zero vector $x \in V$ satisfying $\Omega(x, x) = 0$ is said to be *isotropic*. A linear subspace S is called *totally isotropic* if the restriction of Ω to S is total zero, in other words, S is contained in $S^\perp$. The Grassmannian formed by all k-dimensional totally isotropic subspaces will be denoted by $\mathcal{G}_k(\Omega)$.

From this moment we suppose that the form Ω is non-degenerated. Let $u : V \to V^*$ be the semilinear isomorphism associated with Ω. Then

$$\Omega(x,y) = u(y) \cdot x \quad \forall\, x, y \in V.$$

Since the orthogonal relation is symmetric, we have

$$S^\perp = u(S)^0 = u^{-1}(S^0) \tag{1.6}$$

for every linear subspace $S \subset V$; in particular,

$$\dim S^\perp = \operatorname{codim} S.$$

This implies that $\dim S^{\perp\perp} = \dim S$; since $S \subset S^{\perp\perp}$, we get

$$S^{\perp\perp} = S.$$

If S is totally isotropic then $S \subset S^\perp$ and $\dim S \le \operatorname{codim} S$.

The mapping $S \to S^\perp$ is a bijective transformation of $\mathcal{G}(V)$ sending $\mathcal{G}_k(V)$ to $\mathcal{G}_{n-k}(V)$. This transformation is order reversing:

$$S \subset U \iff U^\perp \subset S^\perp.$$

The square of the transformation is identity.

Remark 1.6. Taking $S^0 = U$ in (1.6) we obtain $u(U^0)^0 = u^{-1}(U)$ which implies that

$$\check{u}(U) = u^{-1}(U).$$

Therefore, $(\check{u})_k = (u^{-1})_k$ for all k. This means that $\check{u}$ is a scalar multiple of u^{-1}. Since $\check{u} = (u^*)^{-1}$, we have $u^* = au$ for non-zero $a \in R$.

1.5.3   *Polarities*

Let V be as in the previous section. A bijection $\pi : \mathcal{G}_1(V) \to \mathcal{G}_{n-1}(V)$ is called a *polarity* if

$$P \subset \pi(Q) \iff Q \subset \pi(P) \quad \forall\, P, Q \in \mathcal{G}_1(V). \tag{1.7}$$

Example 1.11. Let be $\perp$ be the orthogonal relation associated with a non-degenerate reflexive form. Then

$$S \subset U^\perp \iff U \subset S^\perp$$

for any $S, U \in \mathcal{G}(V)$ (since the orthogonal relation is symmetric). This means that the restriction of the transformation $S \to S^\perp$ to $\mathcal{G}_1(V)$ is a polarity.

So, every non-degenerate reflexive form defines a polarity. Conversely, we have the following.

Proposition 1.13. *For every polarity* $\pi : \mathcal{G}_1(V) \to \mathcal{G}_{n-1}(V)$ *there exists a non-degenerate reflexive form* Ω *such that the restriction of the transformation* $S \to S^\perp$ *(where* $\perp$ *is the orthogonal relation associated with* Ω*) to* $\mathcal{G}_1(V)$ *coincides with* π.

Proof. Show that π is a collineation of Π_V to Π_V^*. Let P_1, P_2 be distinct elements of $\mathcal{G}_1(V)$ and $\mathcal{X}$ be the set formed by all $(n-1)$-dimensional linear subspaces containing $P_1 + P_2$. By (1.7), our polarity transfers

$$\mathcal{G}_1(\pi(P_1) \cap \pi(P_2)) \tag{1.8}$$

to the set $\mathcal{X}$. Clearly, $P \in \mathcal{G}_1(V)$ belongs to the line $\mathcal{G}_1(P_1 + P_2)$ if and only if it is contained in every element of $\mathcal{X}$. By (1.7), the latter is equivalent to the fact that $\pi(P)$ contains every element of the set (1.8). Thus

$$P \subset P_1 + P_2 \iff \pi(P_1) \cap \pi(P_2) \subset \pi(P)$$

and π is a collineation.

The Fundamental Theorem of Projective Geometry guarantees the existence of a semilinear isomorphism $u : V \to V^*$ such that $\pi(P) = u(P)^0$ for every $P \in \mathcal{G}_1(V)$. It follows from (1.7) that the sesquilinear form defined by u is reflexive. $\qquad\square$

Chapter 2

Buildings and Grassmannians

This chapter is a survey of basic facts concerning buildings and related topics. Buildings of classical groups will be considered as examples. We introduce the concepts of building Grassmannians and the associated Grassmann spaces. Grassmannians of finite-dimensional vector spaces and Grassmannians formed by totally isotropic subspaces of non-degenerate reflexive forms are special cases of this general construction.

In the second part of the chapter, we prove Abramenko–Van Maldeghem's result on apartments preserving mappings of the chamber sets of buildings (maximal simplices of a building are called chambers). Also we show that the same method cannot be applied to apartments preserving mappings of building Grassmannians.

2.1 Simplicial complexes

2.1.1 *Definition and examples*

Let X be a non-empty set and Δ be a set consisting of *finite* subsets of X (we do not require that X is finite). Suppose that the following conditions hold:

(1) every one-element subset belongs to Δ,
(2) if $A \in \Delta$ then every subset of A belongs to Δ.

Then Δ is called a *simplicial complex*; elements of X and Δ are said to be *vertices* and *simplices*, respectively.

Let Y be a subset of X and Σ be a simplicial complex whose vertex set is Y. If $\Sigma \subset \Delta$ then Σ is called a *subcomplex* of Δ.

Let Δ and Δ' be simplicial complexes whose vertex sets are X and X', respectively. A mapping $f : X \to X'$ is said to be a *morphism* of Δ to Δ' if $f(\Delta) \subset \Delta'$ (the image of every simplex is a simplex). A bijective morphism $f : X \to X'$ is an *isomorphism* if $f(\Delta) = \Delta'$. Isomorphisms of a simplicial complex to itself are called *automorphisms*.

Suppose that X is a set with an incidence relation $*$ (this is a symmetric and reflexive binary relation). A subset consisting of pairwise incident elements is said to be a *flag*. The *flag complex* associated with the relation $*$ is the simplicial complex whose vertex set is X and whose simplices are finite flags.

Now we give a few examples.

Example 2.1. Consider the set of all proper subsets of $\{1, \ldots, n+1\}$ with the natural incidence relation (two subsets are incident if one of them is contained in the other). The associated flag complex is denoted by A_n.

Example 2.2. Let us consider the $2n$-element set

$$J := \{1, \ldots, n, -1, \ldots, -n\}.$$

A subset $X \subset J$ is said to be *singular* if

$$j \in X \implies -j \notin X.$$

We write $\mathcal{J}$ for the set of all singular subsets. Every maximal singular subset consists of n elements and for every $i \in \{1, \ldots, n\}$ it contains i or $-i$. For every $k \in \{1, \ldots, n\}$ we denote by $\mathcal{J}_k$ the set of all singular subset consisting of k elements (every one-element subset is singular and we identify $\mathcal{J}_1$ with J). The flag complex associated with the natural incidence relation on $\mathcal{J}$ will be denoted by C_n.

Now define

$$X_+ := \{1, \ldots, n\}, \quad X_- := \{1, \ldots, n-1, -n\}$$

and consider the sets

$$\mathcal{J}_+ := \{\, X \in \mathcal{J}_n \ : \ n - |X_+ \cap X| \text{ is even }\},$$

$$\mathcal{J}_- := \mathcal{J}_n \setminus \mathcal{J}_+ = \{\, Y \in \mathcal{J}_n \ : \ n - |X_+ \cap Y| \text{ is odd }\}.$$

Then X_+ and X_- are elements of $\mathcal{J}_+$ and $\mathcal{J}_-$, respectively; moreover, $Y \in \mathcal{J}_n$ is an element of $\mathcal{J}_-$ if and only if $n - |X_- \cap Y|$ is even. Also note the following remarkable property: for $X, Y \in \mathcal{J}_n$ the number $n - |X \cap Y|$

is odd if and only if one of these subsets belongs to $\mathcal{J}_+$ and the other is an element of $\mathcal{J}_-$. Let

$$\mathcal{J}^* := \mathcal{J} \setminus \mathcal{J}_{n-1}.$$

We define an incidence relation $*$ on $\mathcal{J}^*$ as follows: in the case when $X \in \mathcal{J}_k$, $k \leq n - 2$, and $Y \in \mathcal{J}^*$, we write $X * Y$ if X and Y are incident in the usual sense (one of these subsets is contained in the other); if $X \in \mathcal{J}_+$ and $Y \in \mathcal{J}_-$ then $X * Y$ means that $X \cap Y$ belongs to $\mathcal{J}_{n-1}$ (for example, X_+ and X_- are incident). The associated flag complex is called the *oriflamme* complex and denoted by D_n.

Example 2.3. Let V be a finite-dimension vector space over a division ring and $\Delta(V)$ be the flag complex defined by the natural incidence relation on the set of all proper linear subspaces of V. If Ω is a non-degenerate reflexive form on V then we write $\Delta(\Omega)$ for the subcomplex of $\Delta(V)$ consisting of all flags formed by totally isotropic subspaces of Ω.

Now suppose that V is a $(2n)$-dimensional vector space over a field whose characteristic is not equal to 2 and Ω is a non-degenerate symmetric form on V. The dimension of maximal totally isotropic subspaces is assumed to be equal to n. Then every $(n - 1)$-dimensional totally isotropic subspace is contained in precisely two maximal totally isotropic subspaces. The action of the group $\mathrm{SO}(\Omega)$ (the group formed by all linear automorphisms of determinant 1 which preserve the form Ω) on the Grassmannian $\mathcal{G}_n(\Omega)$ is not transitive: there are precisely two orbits which will be denoted by $\mathcal{G}_+(\Omega)$ and $\mathcal{G}_-(\Omega)$. Note that for two maximal totally isotropic subspaces S and U the codimension of $S \cap U$ in S is odd if and only if one of these subspaces belongs to $\mathcal{G}_+(\Omega)$ and the other is an element of $\mathcal{G}_-(\Omega)$. Let $\mathcal{G}^*(\Omega)$ be the set of all totaly isotropic subspaces whose dimension is not equal to $n - 1$. As in the previous example, we define the *oriflamme* incidence relation $*$ on $\mathcal{G}^*(\Omega)$. If $S \in \mathcal{G}_k(\Omega)$, $k \leq n - 2$, and $U \in \mathcal{G}^*(\Omega)$ then $S * U$ means that S and U are incident in the usual sense; for $S \in \mathcal{G}_+(\Omega)$ and $U \in \mathcal{G}_-(\Omega)$ we write $S * U$ if their intersection is $(n - 1)$-dimensional. The associated flag complex is called the *oriflamme* complex of Ω and denoted by $\mathrm{Orif}(\Omega)$.

We will restrict ourselves to simplicial complexes satisfying the following condition: every simplex is contained in a certain maximal simplex and all maximal simplices have the same cardinality. The *rank* of a such simplicial complex is the number of vertices in maximal simplices. The flag complexes considered in Examples 2.1, 2.2 and 2.3 satisfy this condition.

2.1.2   *Chamber complexes*

Let Δ be a simplicial complex of rank n. Two maximal simplices C and C' are said to be *adjacent* if

$$|C \cap C'| = n - 1.$$

Let $\mathrm{Ch}(\Delta)$ be the set of all maximal simplices of Δ and $\Gamma_{\mathrm{ch}}(\Delta)$ be the graph whose vertex set is $\mathrm{Ch}(\Delta)$ and whose edges are pairs of adjacent elements. We say that Δ is a *chamber complex* if the graph $\Gamma_{\mathrm{ch}}(\Delta)$ is connected. In this case, maximal simplices of Δ are said to be *chambers*; and we define the *distance* $d(C, C')$ between two chambers C and C' as the distance between the corresponding vertices in the graph $\Gamma_{\mathrm{ch}}(\Delta)$.

In a chamber complex of rank n, every simplex consisting of $n - 1$ vertices is called a *panel*. A chamber complex is said to be *thick* if every panel is contained in at least 3 chambers; and it is called *thin* in the case when every panel is contained in precisely 2 chambers.

Lemma 2.1. *If f and g are isomorphisms between thin chamber complexes Δ and Δ' such that the restrictions of f and g to a certain chamber are coincident then $f = g$.*

Proof.   Suppose that the restrictions of f and g to a chamber C_0 are coincident and consider a chamber C adjacent with C_0. Then $f(C)$ and $g(C)$ are chambers containing the panel

$$f(C_0 \cap C) = g(C_0 \cap C)$$

and distinct from the chamber $f(C_0) = g(C_0)$. Thus $f(C) = g(C)$ (since our chamber complexes are thin) and the restrictions of f and g to C are coincident. By connectedness, the same holds for every chamber of Δ. □

Let Δ be a simplicial complex whose vertex set is X. We say that Δ is *labeled* by a set S if there exists a mapping $\alpha : X \to S$ which satisfies the following conditions:

- the restriction of α to every maximal simplex is bijective,
- for all $A, B \in \Delta$ the inclusion $A \subset B$ implies that $\alpha(A) \subset \alpha(B)$;

in this case, the mapping α is called a *labeling* of Δ. If α is a labeling of Δ by S then for any permutation t on the set S the mapping $t\alpha$ also is a labeling of Δ.

Lemma 2.2. *If Δ is a chamber complex labeled by a set S then any two labelings of Δ by S are coincident up to a permutation on S.*

Proof. Let α and β be labelings of Δ by S. Denote by α' and β' their restrictions to a chamber C_0. Since α' and β' are bijective, there exists a permutation t on S such that $\beta' = t\alpha'$. Show that $\beta = t\alpha$.

We take any chamber C adjacent with C_0. Then $C \setminus C_0$ and $C_0 \setminus C$ are one-point sets. Suppose that

$$C_0 \setminus C = \{x\} \quad \text{and} \quad C \setminus C_0 = \{y\}.$$

Then

$$\alpha(x), \alpha(y) \in S \setminus \alpha(C \cap C_0).$$

Since the latter subset consists of unique element, we get $\alpha(x) = \alpha(y)$. The same holds for the labeling β. The equality $\beta(x) = t\alpha(x)$ (recall that $x \in C_0$) implies that $\beta(y) = t\alpha(y)$. Therefore, the restrictions of β and $t\alpha$ to C are coincident. The connectedness of $\Gamma_{\mathrm{ch}}(\Delta)$ gives the claim. $\square$

Lemma 2.3. *Let Δ be a labeled chamber complex. If 3 distinct chambers of Δ are mutually adjacent then their intersection is a panel.*

Proof. Let α be a labeling of Δ. If C_1 and C_2 are adjacent chambers then

$$C_1 \setminus C_2 = \{x\} \quad \text{and} \quad C_2 \setminus C_1 = \{y\}$$

for certain vertices x, y and $\alpha(x) = \alpha(y)$ (see the proof of the previous lemma). If a chamber C_3 is adjacent with C_1, C_2 and it does not contain the panel $C_1 \cap C_2$ then $x, y \in C_3$; since $\alpha(x) = \alpha(y)$, this contradicts the fact that the restriction of α to C_3 is bijective. Therefore, $C_1 \cap C_2$ is contained in C_3. $\square$

2.1.3 Grassmannians and Grassmann spaces

Let Δ be a chamber complex of rank n and α be a labeling of Δ by a set S. Then $|S| = n$. For every $s \in S$ we define the (α, s)-*Grassmannian*

$$\mathcal{G}_{\alpha,s}(\Delta) := \alpha^{-1}(s).$$

If α' is another labeling of Δ by the set S then, by Lemma 2.2, there exists a permutation t on the set S such that $\alpha' = t\alpha$ and every Grassmannian $\mathcal{G}_{\alpha',s'}(\Delta)$ coincides with a certain Grassmannian $\mathcal{G}_{\alpha,s}(\Delta)$.

Therefore, the vertex set of Δ can be decomposed in n disjoint subsets which will be called *Grassmannians*. This decomposition does not depend on a labeling of the complex Δ.

Let $\mathcal{G}$ be a Grassmannian of Δ. Two distinct vertices $a, b \in \mathcal{G}$ are said to be *adjacent* if there exist adjacent chambers A and B such that $a \in A$ and $b \in B$; this is equivalent to the existence of a panel P such that $P \cup \{a\}$ and $P \cup \{b\}$ are chambers.

The *Grassmann graph* $\Gamma_{\mathcal{G}}$ associated with the Grassmannian $\mathcal{G}$ is the graph whose vertex set is $\mathcal{G}$ and whose edges are pairs of adjacent vertices.

Proposition 2.1. *The Grassmann graph $\Gamma_{\mathcal{G}}$ is connected.*

Proof. Let $a, b \in \mathcal{G}$. We take chambers A and B such that $a \in A$ and $b \in B$. Every path in $\Gamma_{\mathrm{ch}}(\Delta)$ connecting A and B induces a path in $\Gamma_{\mathcal{G}}$ which connects a and b. $\qquad\square$

Every simplex of Δ intersects $\mathcal{G}$ in at most one vertex. Consider a panel P which does not intersect $\mathcal{G}$. The set consisting of all $x \in \mathcal{G}$ such that $P \cup \{x\}$ is a chamber will be called the *line* of $\mathcal{G}$ associated with (defined by) the panel P. Clearly, any two distinct elements of this line are adjacent. Denote by $\mathcal{L}$ the set of all such lines. The partial linear space $\mathfrak{G} := (\mathcal{G}, \mathcal{L})$ is known as the *Grassmann space* or the *shadow space* corresponding to the Grassmannian $\mathcal{G}$. The collinearity relation of $\mathfrak{G}$ coincides with the adjacency relation and the associated collinearity graph is the Grassmann graph $\Gamma_{\mathcal{G}}$.

Let f be an isomorphism of Δ to a chamber complex Δ'. Then Δ' is labeled and f transfers Grassmannians to Grassmannians (if α is a labeling of Δ then αf^{-1} is a labeling of Δ'); moreover, f induces collineations between the Grassmann spaces.

The term "Grassmannian" is motivated by the following example.

Example 2.4. Let V be an n-dimensional vector space over a division ring. The flag complex $\Delta(V)$ is labeled by the dimension function and its rank is equal to $n-1$. This is a chamber complex (it will be shown later). The Grassmannians of $\Delta(V)$ coincide with the Grassmannians of the vector space V. If $k = 1, n-1$ then any two distinct elements of $\mathcal{G}_k(V)$ are adjacent and the associated Grassmann spaces are the projective space Π_V and the dual projective space Π_V^*. In the case when $1 < k < n - 1$, two elements of $\mathcal{G}_k(V)$ are adjacent if their intersection is $(k-1)$-dimensional (this is equivalent to the fact that the sum of these linear subspaces is $(k+1)$-dimensional); every line consists of all elements of $\mathcal{G}_k(V)$ "lying between" two incident linear subspaces of dimension $k-1$ and $k+1$.

2.2 Coxeter systems and Coxeter complexes

2.2.1 *Coxeter systems*

Let W be a group generated by a set S. If $s, s' \in S$ then we write $m(s, s')$ for the order of the element ss'. We suppose that $m(s, s) = 1$ for every $s \in S$ and $m(s, s') \geq 2$ if $s \neq s'$. The first condition guarantees that every element of S is an involution. We have $m(s, s') = m(s', s)$ and the equality $m(s, s') = 2$ holds if and only if the involutions s and s' commute. Denote by $\mathfrak{J}$ the set of all pairs $(s, s') \in S \times S$ such that $m(s, s')$ is finite.

The pair (W, S) is called a *Coxeter system* if the group W has the presentation

$$\langle\, S\ :\ (ss')^{m(s,s')} = 1,\ (s, s') \in \mathfrak{J}\, \rangle;$$

this means that every mapping h of the set S to a group G satisfying

$$(h(s)h(s'))^{m(s,s')} = 1 \quad \forall\, (s, s') \in \mathfrak{J}$$

can be extended to a homomorphism of W to G.

The *diagram* associated with the Coxeter system (W, S) is the graph whose vertex set is S and any two distinct $s, s' \in S$ are connected by $m(s, s') - 2$ edges (in the case when $m(s, s') = \infty$, we draw one edge labeled by ∞). Note that two generators are not connected by an edge if they commute.

Coxeter systems (W, S) and (W', S') are called *isomorphic* if there exists an isomorphism $h : W \to W'$ such that $h(S) = S'$. Coxeter systems are isomorphic if and only if they have the same diagram.

A Coxeter system is called *irreducible* if its diagram is connected.

Remark 2.1. Let $\{S_i\}_{i \in I}$ be the connected components of the diagram associated with a Coxeter system (W, S). We define $W_i := \langle S_i \rangle$ for every $i \in I$. Then each (W_i, S_i) is an irreducible Coxeter system and W is the "bounded" direct product of all W_i, Subsection IV.1.9 in [Bourbaki (1968)].

A Coxeter system (W, S) is called *finite* if W is finite (the group W does not need to be finite if S is finite). There are precisely 11 types of finite irreducible Coxeter systems (Subsection VI.4.1 in [Bourbaki (1968)]):

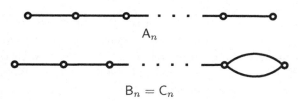

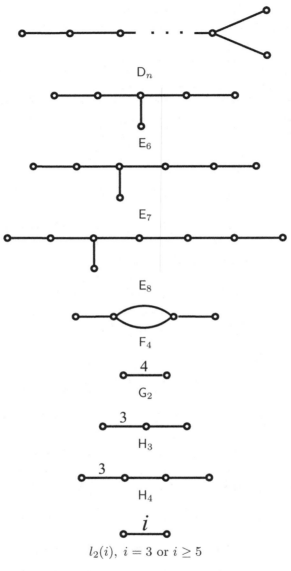

$l_2(i)$, $i = 3$ or $i \geq 5$

Let W be a group generated by a set S. The *length* $l(w) = l_S(w)$ of an element $w \in W$ (with respect to the generating set S) is the smallest number n such that w has an expression $w = s_1 \ldots s_n$ with each $s_i \in S$. An expression $w = s_1 \ldots s_n$ is called *reduced* if $n = l(w)$. We say that the pair (W, S) satisfies the *exchange condition* if for every reduced expression

$w = s_1 \dots s_n$ and every $s \in S$ satisfying $l(sw) \leq n$ there exists $i \in \{1, \dots, n\}$ such that

$$sw = s_1 \dots \hat{s}_i \dots s_n$$

(the symbol ˆ means that the corresponding term is omitted).

Theorem 2.1. *Let W be a group and $S \subset W$ be a set of involutions which generates W. Then (W, S) satisfies the exchange condition if and only if it is a Coxeter system.*

Proof. See Subsection IV.1.6 in [Bourbaki (1968)]. □

The following properties of a Coxeter system (W, S) can be drawn from the exchange condition.

Exercise 2.1. Show that if $s_1, \dots, s_n, s'_1, \dots, s'_n \in S$, $w = s_1 \dots s_n = s'_1 \dots s'_n$ and $l(w) = n$ then

$$\{s_1, \dots, s_n\} = \{s'_1, \dots, s'_n\}.$$

Hint: the proof is induction by $l(w)$; apply the exchange condition to $s_1 w$ and $s'_1 w$.

By Exercise 2.1, for every $w \in W$ there exists a subset $S_w \subset S$ such that every reduced expression of w is formed by all elements of S_w. For every subset $X \subset S$ we define $W_X := \langle X \rangle$.

Exercise 2.2. Show that W_X consists of all $w \in W$ satisfying $S_w \subset X$. *Hint:* establish that

$$S_{sw'} \subset \{s\} \cup S_{w'}$$

for every $s \in S$ and prove that

$$S_{ww'} \subset S_w \cup S_{w'}$$

induction by $l(w)$.

Using Theorem 2.1 and Exercise 2.2 we establish the following.

Theorem 2.2. *Let (W, S) be a Coxeter system. Then for every subset $X \subset S$ the pair (W_X, X) is a Coxeter system, and*

$$W_X \cap W_Y = W_{X \cap Y}$$

for all subsets $X, Y \subset S$.

Proof. See Subsection IV.1.8 in [Bourbaki (1968)]. □

2.2.2   Coxeter complexes

Let (W, S), $S = \{s_1, \ldots, s_n\}$ be a Coxeter system. We define

$$W^k := \langle S \setminus \{s_k\}\rangle, \qquad k = 1, \ldots, n;$$

and for every subset $J = \{j_1, \ldots, j_m\} \subset \{1, \ldots, n\}$

$$W^J := \langle S \setminus \{s_{j_1}, \ldots, s_{j_m}\}\rangle.$$

Then, by Theorem 2.2,

$$W^J = W^{j_1} \cap \cdots \cap W^{j_m}.$$

Exercise 2.3. Show that

$$wW^i = w'W^j \implies i = j \text{ and } ww'^{-1} \in W^i.$$

Thus $wW^i = W^i$ implies that $w \in W^i$, and we have $w = 1$ if the latter equality holds for all i.

The *Coxeter complex* $\Sigma(W, S)$ is the simplicial complex whose vertex set consists of all special subsets wW^k with $w \in W$ and $k \in \{1, \ldots, n\}$; special subsets $X_1, \ldots, X_m$ form a simplex if there exists $w \in W$ such that

$$X_1 = wW^{j_1}, \ldots, X_m = wW^{j_m}.$$

We identify this simplex with the special subset

$$X_1 \cap \cdots \cap X_m = wW^J,$$

where $J = \{j_1, \ldots, j_m\}$. Then every maximal simplex

$$\{wW^1, \ldots, wW^n\}$$

will be identified with the element $w \in W$.

The complex $\Sigma(W, S)$ is finite if the Coxeter system is finite.

Proposition 2.2. *The Coxeter complex* $\Sigma(W, S)$ *is a thin chamber complex.*

Proof.   Denote by C_0 the maximal simplex formed by $W^1, \ldots, W^n$. Then wC_0 is the maximal simplex corresponding to $w \in W$. If $w = s_{i_1} \ldots s_{i_k}$ then

$$C_0, \quad C_1 = s_{i_1}C_0, \quad C_2 = s_{i_1}s_{i_2}C_0, \quad \ldots, \quad C_k = s_{i_1} \ldots s_{i_k}C_0 = wC_0$$

is a path in $\Gamma_{\mathrm{ch}}(\Delta)$. Therefore, C_0 can be connected with every element of $\mathrm{Ch}(\Delta)$ and the graph $\Gamma_{\mathrm{ch}}(\Delta)$ is connected.

Let P_k be the panel formed by

$$W^1, \ldots, \hat{W}^k, \ldots, W^n$$

(the vertex W^k is omitted). If a chamber wC_0 contains P_k then

$$w \in \bigcap_{i \neq k} W^i = \{1, s_k\}$$

(by Exercise 2.3). Thus P_k is contained only in the chambers C_0 and $s_k C_0$. Then wP_k is contained only in the chambers wC_0 and $ws_k C_0$. $\qquad\square$

The mapping $wW^k \to s_k$ is the canonical labeling of $\Sigma(W, S)$ by the set S (by Lemma 2.2, any other labeling of $\Sigma(W, S)$ by S is the composition of the canonical labeling with a permutation on S). We write $\mathrm{Aut}_0(\Sigma(W, S))$ for the group of all automorphisms of $\Sigma(W, S)$ preserving the canonical labeling. In the general case, there exist automorphisms of $\Sigma(W, S)$ which do not belong to this group (see Example 2.5).

Exercise 2.4. For every $w \in W$ denote by l_w the automorphism of $\Sigma(W, S)$ which sends every special subset X to wX. Show that $w \to l_w$ is an isomorphism of W to the group $\mathrm{Aut}_0(\Sigma(W, S))$. *Hint:* use Lemma 2.1 to prove the surjectivity.

The Grassmannians of $\Sigma(W, S)$ coincide with the orbits of the action of the group $W = \mathrm{Aut}_0(\Sigma(W, S))$ on the vertex set of $\Sigma(W, S)$. Denote by $\mathcal{G}_k(W, S)$ the Grassmannian containing W^k; it consists of all special subsets wW^k, $w \in W$. Grassmannians of Coxeter complexes are closely related with the concept of so-called *Coxeter matroids* [Borovik, Gelfand and White (2003)].

2.2.3 *Three examples*

In this subsection we consider the Coxeter systems of types $\mathsf{A}_n, \mathsf{C}_n, \mathsf{D}_n$ and show that the associated Coxeter complexes are $\mathsf{A}_n, \mathsf{C}_n, \mathsf{D}_n$ (respectively).

Example 2.5 (Type A_n, $n \geq 1$). Let $W = \mathrm{S}_{n+1}$, where S_{n+1} is the symmetric group consisting of all permutations of the set $I := \{1, \ldots, n+1\}$. Let also S be the set formed by the transpositions

$$s_i = (i, i+1), \quad i = 1, \ldots, n.$$

Then (W, S) is a Coxeter system whose diagram is

For every number $k \in \{1, \ldots, n\}$ the subgroup W^k is the stabilizer of the subset $\{1, \ldots, k\}$ and the mapping

$$wW^k \to \{w(1), w(2), \ldots, w(k)\}$$

is well-defined. This is an isomorphism of $\Sigma(W, S)$ to the complex A_n (Example 2.1); it transfers the Grassmannian $\mathcal{G}_k(W, S)$ to the set of all k-element subsets of I. The automorphism of the complex A_n sending every subset $X \subset I$ to $I \setminus X$ induces an automorphism of $\Sigma(W, S)$ which does not preserve the canonical labeling.

Example 2.6 (Type $\mathsf{B}_n = \mathsf{C}_n$, $n \geq 2$). We say that a permutation s on the set $J := \{1, \ldots, n, -1, \ldots, -n\}$ is *symplectic* if

$$s(-j) = -s(j) \quad \forall \, j \in J.$$

All symplectic permutations form the group denoted by Sp_n. Elements of this group preserve $\mathcal{J}$ (the set of all singular subsets, see Example 2.2). The group Sp_n is generated by the "symplectic transpositions"

$$s_i = (i, i+1)(-i, -(i+1)), \quad i = 1, \ldots, n-1,$$

and the transposition

$$s_n = (n, -n).$$

Suppose that $W = \mathrm{Sp}_n$ and S is the set of generators considered above. Then (W, S) is a Coxeter system with the diagram

As in the previous example, the subgroup W^k is the stabilizer of the subset $\{1, \ldots, k\}$ and the mapping

$$wW^k \to \{w(1), w(2), \ldots, w(k)\}$$

is an isomorphism of $\Sigma(W, S)$ to the complex C_n (Example 2.2). This isomorphism sends the Grassmannian $\mathcal{G}_k(W, S)$ to $\mathcal{J}_k$ (the set formed by all singular subsets consisting of k elements).

Example 2.7 (Type D_n, $n \geq 4$). In this example we suppose that W is the subgroup $SO_n \subset Sp_n$ generated by the set S consisting of $s_1, \ldots, s_{n-1}$ (defined in the previous example) and the "symplectic transposition"

$$s_n = (n-1, -n)(-(n-1), n).$$

Then (W, S) is a Coxeter system whose diagram is

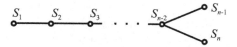

The action of the group W on the set $\mathcal{J}_n$ is not transitive. For example, W does not contain permutations transferring

$$X_+ := \{1, \ldots, n\} \quad \text{to} \quad X_- := \{1, \ldots, n-1, -n\}$$

(since every symplectic permutation sending X_+ to X_- is the composition of $(n, -n)$ and an element of W^n). The reader can show that $\mathcal{J}_+$ and $\mathcal{J}_-$ (Example 2.2) are the orbits of the action of W on $\mathcal{J}_n$ and the mapping

$$wW^k \to \{w(1), w(2), \ldots, w(k)\} \quad \text{if} \quad k \leq n-2,$$

$$wW^{n-1} \to \{w(1), \ldots, w(n-1), w(-n)\},$$

$$wW^n \to \{w(1), \ldots, w(n-1), w(n)\}$$

is an isomorphism of $\Sigma(W, S)$ to the oriflamme complex D_n (Example 2.2). This isomorphism transfers the Grassmannians of $\Sigma(W, S)$ to the sets $\mathcal{J}_1, \ldots, \mathcal{J}_{n-2}, \mathcal{J}_-, \mathcal{J}_+$.

2.3 Buildings

2.3.1 *Definition and elementary properties*

Let Δ be a simplicial complex and $\mathfrak{A}$ be a set of subcomplexes of Δ satisfying the following axioms:

(1) every element of $\mathfrak{A}$ is isomorphic to a Coxeter complex,
(2) for any two simplices of Δ there is an element of $\mathfrak{A}$ containing both of them.

Then Δ is a chamber complex (since every Coxeter complex is a chamber complex, Proposition 2.2). The simplicial complex Δ is a *building* if the following additional axiom holds:

(3) for all $\Sigma, \Sigma' \in \mathfrak{A}$ such that $\Sigma \cap \Sigma'$ contains a chamber of Δ there exists an isomorphism of Σ to Σ' preserving pointwise all simplices contained in $\Sigma \cap \Sigma'$ (by Lemma 2.1, such isomorphism is unique);

in this case, elements of $\mathfrak{A}$ are said to be *apartments* and $\mathfrak{A}$ is called a *system of apartments*.

The axiom (3) guarantees that any two apartments $\Sigma, \Sigma' \in \mathfrak{A}$ are isomorphic: we take chambers $C \in \Sigma$ and $C' \in \Sigma'$ and consider an apartment Σ'' containing them; by the axiom (3), Σ'' is isomorphic to both Σ and Σ'. Similar arguments can be used to prove the following property (we leave the details for the reader):

(4) if $\Sigma, \Sigma' \in \mathfrak{A}$ then for any simplices $A, B \in \Sigma \cap \Sigma'$ there exists an isomorphism of Σ to Σ' preserving A and B pointwise.

Remark 2.2. The axiom systems (1), (2), (3) and (1), (2), (4) are equivalent, Section IV.1 in [Brown (1989)]. Moreover, by Section 4.3 in [Garrett (1997)], the axiom (1) can be drawn from the axioms (2) and (3).

Example 2.8. Every Coxeter complex can be considered as a building with unique apartment.

Let Δ be a building and $\mathfrak{A}$ be a system of apartments for Δ. We write $\mathfrak{A}_C$ for the set of all apartments containing a chamber C. By the axiom (3), for any two apartments $\Sigma, \Sigma' \in \mathfrak{A}_C$ there is a unique isomorphism $f_{\Sigma\Sigma'}$ of Σ to Σ' whose restriction to C is identity. We have

$$f_{\Sigma'\Sigma} = f_{\Sigma''\Sigma} f_{\Sigma'\Sigma''}$$

for all $\Sigma, \Sigma', \Sigma'' \in \mathfrak{A}_C$ (because $f_{\Sigma''\Sigma} f_{\Sigma'\Sigma''}$ is an isomorphism of Σ' to Σ preserving C pointvise). Now we fix an apartment $\Sigma \in \mathfrak{A}_C$. Since the union of all apartments from $\mathfrak{A}_C$ coincides with Δ (by the axiom (2)), all isomorphisms

$$f_{\Sigma'\Sigma}, \quad \Sigma' \in \mathfrak{A}_C,$$

fit together to give a surjective morphism of Δ to Σ. This morphism will be denoted by $\rho_{\Sigma,C}$ and called the *retraction* of Δ on Σ *centered* at C.

If α is a labeling of Σ (apartments are labeled by the axiom (1)) then $\alpha\rho_{\Sigma,C}$ is a labeling of Δ. Therefore, we have the following.

Proposition 2.3. *Every building is a labeled chamber complex.*

Proposition 2.4. *Let $\Sigma \in \mathfrak{A}$ and C, C' be chambers of Σ. Then Σ contains every geodesic of the graph $\Gamma_{\mathrm{ch}}(\Delta)$ connecting C and C'.*

Proof. Suppose that

$$C = C_0, C_1, \ldots, C_k = C'$$

is a geodesic of $\Gamma_{\mathrm{ch}}(\Delta)$ which is not contained in Σ. We choose an index i such that $C_{i-1} \in \Sigma$ and $C_i \notin \Sigma$. There is a unique chamber $C'' \in \Sigma$ distinct from C_{i-1} and containing the panel $C_{i-1} \cap C_i$. Consider the retraction $\rho = \rho_{\Sigma,C''}$. The chamber $\rho(C_i) \in \Sigma$ contains the panel

$$\rho(C_{i-1} \cap C_i) = C_{i-1} \cap C_i,$$

hence it coincides with C_{i-1} or C''. Since there is an apartment containing both C'' and C_i, we have

$$\rho(C_i) \neq \rho(C'') = C''.$$

So $\rho(C_i) = C_{i-1}$ and

$$C = C_0, C_1, \ldots, C_{i-1} = \rho(C_i), \rho(C_{i+1}), \ldots, \rho(C_k) = C'$$

is a part in $\Gamma_{\mathrm{ch}}(\Delta)$ (if A, B are adjacent chambers then $\rho(A), \rho(B)$ are adjacent or coincident). This path contains at most $k - 1$ edges which contradicts the assumption that $d(C, C') = k$. □

In general, the building Δ can admit different systems of apartments; but the union of any collection of apartment systems is again an apartment system, Section IV.4 in [Brown (1989)]. This implies the existence of a largest system of apartments. Moreover, there exists a unique (up to isomorphism) Coxeter system (W, S) such that all apartments of Δ are isomorphic to the Coxeter complex $\Sigma(W, S)$, Section IV.3 in [Brown (1989)].

2.3.2 *Buildings and Tits systems*

Suppose that G is a group spanned by proper subgroups B and N; moreover, $B \cap N$ is a normal subgroup of N. Let S be a set of generators of the quotient group

$$W := N/(B \cap N).$$

If $w \in W$ then for any two elements g_1 and g_2 belonging to the class w we have $g_2^{-1} g_1 \in B \cap N$ and $g_1 B = g_2 B$; thus wB and

$$C(w) := BwB$$

are well-defined. We require that the following two technical conditions hold:

(1) $C(s)C(w) \subset C(w) \cup C(sw)$ for all $s \in S$ and $w \in W$,

(2) sBs^{-1} is not contained in B for every $s \in S$.

Then (G, B, N, S) is said to be a *Tits system* and W is called the *Weyl group* of this Tits system; also we say that the subgroup B, N form a BN-*pair* of the group G.

The conditions (1) and (2) give the following remarkable properties, Section V.2 in [Brown (1989)]:

(3) every $s \in S$ is an involution and (W, S) is a Coxeter system;

(4) $C(w) \cap C(w') = \emptyset$ if $w \neq w'$;

(5) for every subgroup $W' = \langle S' \rangle$ with $S' \subset S$

$$BW'B := \bigcup_{w \in W'} C(w)$$

is a subgroup of G; in particular, we have $BWB = G$, and $BW'B$ coincides with B if $W' = \{1\}$;

(6) $C(s)C(w) = C(sw)$ if $l(sw) \geq l(w)$, and $C(s)C(w) = C(w) \cup C(sw)$ if $l(sw) \leq l(w)$.

Now assume that $S = \{s_1, \ldots, s_n\}$. For every number $k \in \{1, \ldots, n\}$ and every subset $J \subset \{1, \ldots, n\}$ we define W^k and W^J as in Subsection 2.2.2. By the property (5), all

$$P^J := BW^J B, \quad J \subset \{1, \ldots, n\},$$

are subgroup of G and

$$P^J = P^{j_1} \cap \cdots \cap P^{j_m} \quad \text{if} \quad J = \{j_1, \ldots, j_m\}.$$

These subgroups are called *special*. Every special subgroup contains B. Conversely, every subgroup of G containing B is special, Section V.2 in [Brown (1989)].

Exercise 2.5. Show that

$$gP^i = g'P^j \implies i = j \text{ and } gg'^{-1} \in P^i.$$

Consider the simplicial complex $\Delta = \Delta(G, B, N, S)$ whose vertex set consists of all special subsets gP^k with $g \in G$ and $k \in \{1, \ldots, n\}$; special subsets $X_1, \ldots, X_m$ form a simplex if there exists $g \in G$ such that

$$X_1 = gP^{j_1}, \ldots, X_m = gP^{j_m}.$$

We identify this simplex with the special subset

$$X_1 \cap \cdots \cap X_m = gP^J,$$

where $J = \{j_1, \ldots, j_m\}$. Then maximal simplices will be identified with special subsets gB.

Remark 2.3. A subgroup of G is called *parabolic* if it is conjugate to a special subgroup. There is a one-to-one correspondence between simplices of Δ and parabolic subgroups, given by $gP \to gPg^{-1}$ (it easy follows from the fact that there are no two distinct special subgroups which are conjugate).

If $w \in W$ then for any two elements g_1 and g_2 of the class w we have $g_1 P^J = g_2 P^J$ for every $J \subset \{1, \ldots, n\}$ and denote this special subset by wP^J. Let Σ be the subcomplex of Δ formed by all special subsets wP^J with $w \in W$, $J \subset \{1, \ldots, n\}$; it will be called the *fundamental apartment* of Δ. The mapping

$$wW^k \to wP^k$$

is an isomorphism of the Coxeter complex $\Sigma(W, S)$ to the fundamental apartment Σ. For every $g \in G$ denote by l_g the automorphism of Δ transferring every special subset X to gX. The subcomplexes

$$g\Sigma := l_g(\Sigma), \quad g \in G,$$

are called *apartments* of Δ. Then $\{g\Sigma\}_{g \in G}$ is an apartment system and Δ is a building, Section V.3 in [Brown (1989)].

The mapping $gP^k \to s_k$ is the canonical labeling of the building Δ by the set S (any other labeling of Δ by S is the composition of the canonical labeling with a permutation on S, Lemma 2.2). Denote by $\mathrm{Aut}_0(\Delta)$ the group formed by all automorphisms of Δ preserving the canonical labeling (there exist automorphisms of Δ which do not belong to this group, see Remark 2.4). The mapping $g \to l_g$ is a homomorphism of G to $\mathrm{Aut}_0(\Delta)$; it does not need to be injective and surjective (Remark 2.4).

The Grassmannians of Δ coincide with the orbits of the left action of the group G on the vertex set of Δ. Denote by $\mathcal{G}_k(\Delta)$ the Grassmannian containing P^k, it consists of all special subsets gP^k.

2.3.3 Classical examples

In this subsection we consider buildings associated with general linear, symplectic, and orthogonal groups.

Example 2.9 (The general linear group). Let $G = \mathrm{GL}(V)$, where V is an $(n+1)$-dimensional vector space over a division ring. We take any base

$X_0 = \{x_1, \ldots, x_{n+1}\}$ of this vector space and denote by B the stabilizer of the maximal flag

$$\langle x_1 \rangle \subset \langle x_1, x_2 \rangle \subset \cdots \subset \langle x_1, \ldots, x_n \rangle$$

in the group G. Let also $N \subset G$ be the stabilizer of the projective base $\langle x_1 \rangle, \ldots, \langle x_{n+1} \rangle$. Then $B \cap N$ consists of all linear automorphisms preserving each $\langle x_i \rangle$. This is a normal subgroup of N and the corresponding quotient group is isomorphic to the symmetric group S_{n+1}. Let S be the set of generators of S_{n+1} considered in Example 2.5. Then (G, B, N, S) is a Tits system, Section V.5 in [Brown (1989)]. The associated building Δ can be identified with the flag complex $\Delta(V)$ (Example 2.3). Indeed, for every number $k \in \{1, \ldots, n\}$ the subgroup P^k is the stabilizer of the linear subspace $\langle x_1, \ldots, x_k \rangle$ and the mapping

$$gP_k \rightarrow g(\langle x_1, \ldots, x_k \rangle)$$

is the required isomorphism. For every base $X \subset V$ denote by Σ_X the subcomplex of $\Delta(V)$ consisting of all flags formed by the linear subspaces spanned by subsets of X; this subcomplex is isomorphic to the complex A_n (Example 2.1). Then the fundamental apartment Σ is identified with Σ_{X_0} and $g\Sigma = \Sigma_{g(X_0)}$ for every $g \in G$. The Grassmannian $\mathcal{G}_k(\Delta)$ (the Grassmannian containing P^k) is $\mathcal{G}_k(V)$. The associated Grassmann spaces were described in Example 2.4.

Remark 2.4. By the Fundamental Theorem of Projective Geometry (Corollary 1.5), every element of $\mathrm{Aut}_0(\Delta(V))$ is induced by a semilinear automorphism of V. For every homothetic transformation the associated automorphism of $\Delta(V)$ is identity. Hence the homomorphism $g \rightarrow l_g$ of $\mathrm{GL}(V)$ to the group $\mathrm{Aut}_0(\Delta(V))$ is not injective. This homomorphism is surjective only in the case when every automorphism of the associated division ring is inner, for example, if V is a real vector space or a vector space over the division ring of real quaternion numbers (indeed, if l is a σ-linear automorphism of V such that $(l)_1 = (s)_1$ for a certain linear automorphism $s : V \rightarrow V$ then l is a scalar multiple of s which is equivalent to the fact that the automorphism σ is inner). Also note that semilinear isomorphisms of V to V^* (if they exist) induce automorphisms of $\Delta(V)$ which do not preserve the canonical labeling.

Example 2.10 (The symplectic group). Let Ω be a non-degenerate alternating form on a $(2n)$-dimensional vector space V over a field. Then every 1-dimensional linear subspace is totally isotropic and the dimension of maximal totally isotropic subspaces is equal to n. A base $X \subset V$ is said to be an Ω-*base* if for each $x \in X$ there is precisely one $y \in X$ satisfying $\Omega(x, y) \neq 0$. Such bases form a sufficiently wide class.

We fix an Ω-base

$$X_0 = \{x_1, y_1, \ldots, x_n, y_n\}$$

satisfying $\Omega(x_i, y_i) \neq 0$ for all i. Suppose that $G = \mathrm{Sp}(\Omega)$ (the group formed by all linear automorphisms of V preserving the form Ω). Let $B \subset G$ be the stabilizer of the flag

$$\langle x_1 \rangle \subset \langle x_1, x_2 \rangle \subset \cdots \subset \langle x_1, \ldots, x_n \rangle$$

(all linear subspaces in this flag are totally isotropic) and $N \subset G$ be the stabilizer of the projective base

$$\langle x_1 \rangle, \langle y_1 \rangle, \ldots, \langle x_n \rangle, \langle y_n \rangle.$$

Then $B \cap N$ is a normal subgroup of N and the corresponding quotient group is isomorphic to Sp_n. If S is the set of generators of Sp_n considered in Example 2.6 then (G, B, N, S) is a Tits system, Section V.6 in [Brown (1989)]. As in the previous example, we show that the associated building can be identified with the subcomplex $\Delta(\Omega) \subset \Delta(V)$ (Example 2.3). The fundamental apartment is the intersection of $\Delta(\Omega)$ with Σ_{X_0}, and every apartment is the subcomplex $\Delta(\Omega) \cap \Sigma_X$, where X is a certain Ω-base. These apartments are isomorphic to the complex C_n (Example 2.2).

The Grassmannians of this building are $\mathcal{G}_k(\Omega)$, $k \in \{1, \ldots, n\}$. Two elements of $\mathcal{G}_n(\Omega)$ are adjacent if their intersection is $(n-1)$-dimensional; every line in $\mathcal{G}_n(\Omega)$ consists of all maximal totally isotropic subspaces containing a certain element of $\mathcal{G}_{n-1}(\Omega)$. In the case when $k < n$, elements $S, U \in \mathcal{G}_k(\Omega)$ are adjacent if $S \perp U$ and $S \cap U$ belongs to $\mathcal{G}_{k-1}(\Omega)$ (this is equivalent to the fact that $S + U$ is an element of $\mathcal{G}_{k+1}(\Omega)$); the lines are defined as for Grassmannians of finite-dimensional vector spaces (by pairs of incident totally isotropic subspaces of dimension $k - 1$ and $k + 1$).

Example 2.11 (The orthogonal group). Let V be a $(2n)$-dimensional vector space over a field whose characteristic is not equal to 2. Let also Ω be a non-degenerate symmetric form defined on V. The dimension of maximal totally isotropic subspaces is assumed to be equal to n. We say that a base $X \subset V$ is an Ω-*base* if it consists of isotropic vectors and for each $x \in X$ there is precisely one $y \in X$ satisfying $\Omega(x, y) \neq 0$. As in the previous example, these bases form a sufficiently wide class.

Consider the oriflamme complex $\mathrm{Orif}(\Omega)$ (Example 2.3). Every Ω-base X defines the subcomplex $\mathrm{Orif}_X \subset \mathrm{Orif}(\Omega)$ consisting of all "oriflamme flags" formed by the totally isotropic subspaces spanned by subsets of X; this subcomplex is isomorphic to the oriflamme complex D_n (Example 2.2). Let us fix an Ω-base $X_0 = \{x_1, \ldots, x_{2n}\}$ and a maximal simplex (a maximal oriflamme flag) in the subcomplex Orif_{X_0}. Suppose that $G = \mathrm{SO}(\Omega)$ and denote by B the stabilizer of this maximal simplex in G. Let $N \subset G$ be the stabilizer of the projective base $\langle x_1 \rangle, \ldots, \langle x_{2n} \rangle$. Then $B \cap N$ is a normal subgroup of N and the corresponding quotient group is isomorphic to SO_n. Let S be the set of generators of the group SO_n from Example 2.7. Then (G, B, N, S) is a Tits system, Section V.7 in [Brown (1989)]. Standard arguments show that the associated building can be identified with the oriflamme complex $\mathrm{Orif}(\Omega)$. Every apartment of this building is Orif_X, where X is an Ω-base.

The Grassmannians of this building are $\mathcal{G}_k(\Omega)$, $k \leq n - 2$, and $\mathcal{G}_\delta(\Omega)$, $\delta \in \{+, -\}$. The adjacency relation on $\mathcal{G}_k(\Omega)$, $k \leq n-2$, is defined as in the previous example. Two elements of $\mathcal{G}_\delta(\Omega)$ are adjacent if their intersection belongs to $\mathcal{G}_{n-2}(\Omega)$; every line consists of all elements of $\mathcal{G}_\delta(\Omega)$ containing a certain $(n - 2)$-dimensional totally isotropic subspace.

Example 2.12. Let G be a reductive algebraic group over a field and B be a Borel subgroup of G containing a maximal torus T (we refer [Humphreys (1975)] for the precise definitions). Denote by N the normalizer of T in G. Then B and N form a BN-pair of G. The associated Tits system and building are described in [Tits (1974)] (Chapter 5).

2.3.4 *Spherical buildings*

A building Δ is called *spherical* if the associated Coxeter system is finite. In this case, by Proposition 2.4, the diameter of the graph $\Gamma_{\mathrm{ch}}(\Delta)$ is finite and equal to the diameter of the graph $\Gamma_{\mathrm{ch}}(\Sigma)$, where Σ is an apartment of Δ. We say that two chambers of a spherical building are *opposite* if the distance between them is maximal (is equal to the diameter).

The term "spherical" is motivated by the following.

Proposition 2.5. *In a spherical building every apartment Σ is the union of all geodesics connecting two opposite chambers $C, C' \in \Sigma$.*

Proof.   See Section IV.5 in [Brown (1989)]. $\qquad\qquad\qquad\qquad\qquad\square$

Corollary 2.1. *Every spherical building admits a unique system of apartments; in other words, if Δ is a spherical building and $\mathfrak{A}, \mathfrak{A}'$ are apartment systems for Δ then $\mathfrak{A} = \mathfrak{A}'$.*

Corollary 2.2. *Every isomorphism between spherical buildings is apartments preserving (sends apartments to apartments).*

Proof. Let Δ, Δ' be spherical buildings and $\mathfrak{A}, \mathfrak{A}'$ be their systems of apartments. If f is an isomorphism of Δ to Δ' then $f(\mathfrak{A})$ is an apartment system for Δ'. By Corollary 2.1, we have $f(\mathfrak{A}) = \mathfrak{A}'$. $\square$

A building is called *irreducible* if the associated Coxeter system is irreducible. Irreducible thick spherical buildings of rank ≥ 3 were classified in [Tits (1974)]. There are three *classical* types

$$\mathsf{A}_n, \quad \mathsf{B}_n = \mathsf{C}_n, \quad \mathsf{D}_n,$$

and four *exceptional* types

$$\mathsf{F}_4 \quad \text{and} \quad \mathsf{E}_i, \; i = 6, 7, 8$$

(the building type is the type of the associated Coxter system). In particular, there exists no thick building of type H_i, $i = 3, 4$.

We restrict ourselves to buildings of classical types only. Every thick building of type A_n, $n \geq 3$, is isomorphic to the flag complex $\Delta(V)$, where V is an $(n + 1)$-dimensional vector space over a division ring (this fact is a reformulation of Theorem 1.3). All thick buildings of types C_n and D_n can be obtained from so-called *polar spaces* (this construction will be considered in Chapter 4).

2.3.5 *Mappings of the chamber sets*

Let Δ and Δ' be buildings. Isomorphisms of Δ to Δ' (if they exist) induce isomorphisms between the graphs $\Gamma_{\mathrm{ch}}(\Delta)$ and $\Gamma_{\mathrm{ch}}(\Delta')$. Conversely, we have the following.

Theorem 2.3 (J. Tits). *If the diagram of the Coxeter system associated with Δ does not contain edges labeled by ∞ then every isomorphism of $\Gamma_{\mathrm{ch}}(\Delta)$ to $\Gamma_{\mathrm{ch}}(\Delta')$ is induced by an isomorphism of Δ to Δ'.*

Proof. See [Tits (1974)], p.51. $\square$

Theorem 2.4 ([Abramenko and Van Maldeghem (2000)]). *If Δ is a thick spherical building then for any two distinct chambers $C_1, C_2 \in \Delta$ the following conditions are equivalent:*

- C_1 and C_2 are adjacent,
- there exists a chamber $C \neq C_1, C_2$ such that no chamber is opposite to a unique member of the set $\{C, C_1, C_2\}$.

In particular, if Δ' also is a thick spherical building then every bijection of $\mathrm{Ch}(\Delta)$ to $\mathrm{Ch}(\Delta')$ preserving the relation to be opposite (two chambers of Δ are opposite if and only if their images are opposite) is an isomorphism of $\Gamma_{\mathrm{ch}}(\Delta)$ to $\Gamma_{\mathrm{ch}}(\Delta')$.

The intersection of an apartment of Δ with the chamber set $\mathrm{Ch}(\Delta)$ will be called an *apartment* in $\mathrm{Ch}(\Delta)$. We say that a mapping

$$f : \mathrm{Ch}(\Delta) \to \mathrm{Ch}(\Delta')$$

is *apartments preserving* if its restriction to every apartment of $\mathrm{Ch}(\Delta)$ is a bijection on a certain apartment of $\mathrm{Ch}(\Delta')$. Every apartments preserving mapping is injective (since for any two chambers there is an apartment containing them). We want to show that apartments preserving mappings are adjacency preserving; our proof will be based on the following characterization of the adjacency relation in terms of apartments.

Lemma 2.4. *Let Δ be a thick building. Distinct chambers $C, C' \in \Delta$ are adjacent if and only if the intersection of all apartments of $\mathrm{Ch}(\Delta)$ containing them coincides with $\{C, C'\}$.*

Proof. If the intersection of all apartments of $\mathrm{Ch}(\Delta)$ containing both C and C' coincides with $\{C, C'\}$ then, by Proposition 2.4, C and C' are adjacent.

Conversely, suppose that C and C' are adjacent and denote by $\mathcal{X}$ the intersection of all apartments of $\mathrm{Ch}(\Delta)$ containing C, C'. Assume that $\mathcal{X} \neq \{C, C'\}$.

First we establish the existence of $C'' \in \mathcal{X} \setminus \{C, C'\}$ such that

$$C, C', C'' \quad \text{or} \quad C'', C, C'$$

is a geodesic in $\Gamma_{\mathrm{ch}}(\Delta)$. We take any geodesic γ of $\Gamma_{\mathrm{ch}}(\Delta)$ connecting C with a certain element of $\mathcal{X} \setminus \{C, C'\}$; by Proposition 2.4, γ is contained in $\mathcal{X}$. If γ contains C' then the first possibility is realized. If C' does not belong to γ then γ contains a chamber C'' ($C'' \neq C'$) adjacent with C; this chamber is not adjacent with C' (otherwise, by Lemma 2.3, C'' contains the panel $C \cap C'$ which is impossible, since $C, C', C'' \in \mathcal{X}$ and apartments are thin chamber complexes).

Consider the first case $(C, C', C''$ is a geodesic); the second is similar. We choose a chamber $\hat{C}$ containing the panel $C' \cap C''$ and distinct from C' and C'' (our building is thick). The chambers C and $\hat{C}$ are not adjacent (it follows from Lemma 2.3). Hence $C, C', \hat{C}$ is a geodesic in $\Gamma_{\mathrm{ch}}(\Delta)$. Then every apartment containing C and $\hat{C}$ contains C'; but C'' does not belong to this apartment which contradicts $C'' \in \mathcal{X}$. $\qquad\square$

Theorem 2.5. *Suppose that Δ and Δ' are thick buildings of the same rank. Then every apartments preserving mapping of $\mathrm{Ch}(\Delta)$ to $\mathrm{Ch}(\Delta')$ preserves the adjacency relation (two chambers of Δ are adjacent if and only if their images are adjacent); in particular, every apartments preserving bijection of $\mathrm{Ch}(\Delta)$ to $\mathrm{Ch}(\Delta')$ is an isomorphism of $\Gamma_{\mathrm{ch}}(\Delta)$ to $\Gamma_{\mathrm{ch}}(\Delta')$.*

Proof. Let $f : \mathrm{Ch}(\Delta) \to \mathrm{Ch}(\Delta')$ be an apartments preserving mapping. It was noted above that f is injective. Lemma 2.4 guarantees that f transfers adjacent chambers to adjacent chambers.

Let C and $\hat{C}$ be chambers of Δ and $\mathcal{A}$ be an apartment of $\mathrm{Ch}(\Delta)$ containing them. Suppose that the rank of Δ is equal to n. Then $\mathcal{A}$ contains precisely n distinct chambers adjacent with C; denote them by $C_1, \ldots, C_n$. Their images

$$f(C_1), \ldots, f(C_n) \in f(\mathcal{A})$$

are adjacent with $f(C)$. Since the rank of Δ' also is equal to n, the apartment $f(\mathcal{A})$ does not contain other chambers adjacent with $f(C)$. Hence if $f(C)$ and $f(\hat{C})$ are adjacent then we have $\hat{C} = C_i$ for a certain i. Therefore, C and $\hat{C}$ are adjacent if and only if their images are adjacent. $\qquad\square$

Remark 2.5. Theorem 2.5 is a modification of a result obtained in [Abramenko and Van Maldeghem (2008)]. We do not assume that the mapping is bijective, but require that the buildings are of the same rank.

2.4 Mappings of Grassmannians

The intersections of a building Grassmannian with apartments of the associated building are called *apartments* of this Grassmannian. We say that a mapping between two building Grassmannians is *apartments preserving* if its restriction to every apartment is a bijection to an apartment. Apartments preserving mappings are injective, since for any two elements of a building Grassmannian there is an apartment containing them.

Example 2.13. Let V be an n-dimensional vector space over a division ring. Every apartment of the Grassmannian $\mathcal{G}_k(V)$ consists of all k-dimensional linear subspaces spanned by subsets of a certain base of V. In the cases when $k = 1, n - 1$, this is a base of the projective space Π_V or the dual projective space Π_V^*, respectively.

Let V and V' be vector spaces of the same finite dimension $n \geq 3$. It is not difficult to prove that every apartments preserving bijection of $\mathcal{G}_1(V)$ to $\mathcal{G}_1(V')$ (a bijection which sends bases of Π_V to bases of $\Pi_{V'}$) is a collineation of Π_V to $\Pi_{V'}$; similarly, every apartments preserving bijection of $\mathcal{G}_{n-1}(V)$ to $\mathcal{G}_{n-1}(V')$ is a collineation of Π_V^* to $\Pi_{V'}^*$. In Chapters 3 and 4 we establish such kind results for other Grassmannians associated with buildings of classical types.

Let Δ and Δ' be buildings of the same classical type X_n, $\mathsf{X} \in \{\mathsf{A}, \mathsf{C}, \mathsf{D}\}$. Let also $\mathcal{G}$ and $\mathcal{G}'$ be Grassmannians of Δ and Δ', respectively. Denote by $\mathfrak{G}$ and $\mathfrak{G}'$ the associated Grassmann spaces (Subsection 2.1.3).

We suppose that $\mathfrak{G}$ and $\mathfrak{G}'$ both are not projective spaces. In this case, we show that every isomorphism of $\Gamma_\mathcal{G}$ to $\Gamma_{\mathcal{G}'}$ (Subsection 2.1.3) is a collineation of $\mathfrak{G}$ to $\mathfrak{G}'$. For projective spaces this fails (since any two points of a projective space are collinear and every bijection between projective spaces gives an isomorphism of their collinearity graphs).

We describe all collineations of $\mathfrak{G}$ to $\mathfrak{G}'$ and show that they can be extended to isomorphisms of Δ to Δ' (for projective spaces the latter statement is trivial). In almost all cases (except the case when $\mathsf{X} \in \{\mathsf{C}, \mathsf{D}\}$ and $n = 4$), these results easy follow from elementary properties of maximal singular subspaces of Grassmann spaces. Note that such kind theorems were first proved in [Chow (1949)] for Grassmannians of finite-dimensional vector spaces and Grassmannians formed by maximal totally isotropic subspaces of non-degenerate reflexive forms.

One of our main results is the description of all apartments preserving mappings of $\mathcal{G}$ to $\mathcal{G}'$; in particular, we establish that every apartments preserving bijection of $\mathcal{G}$ to $\mathcal{G}'$ is a collineation of $\mathfrak{G}$ to $\mathfrak{G}'$ (hence it can be extended to an isomorphism of Δ to Δ').

The idea used to prove Theorem 2.5 does not work for building Grassmannians, since for any $a, b \in \mathcal{G}$ the intersection of all apartments of $\mathcal{G}$ containing both a, b coincides with $\{a, b\}$.

Our method is based on combinatorial properties of so-called *maximal inexact subsets*. We say that a subset of an apartment is *inexact* if there exist other apartments containing this subset. Certain Grassmannians have maximal inexact subsets of precisely one type; but there exist other Grassmannians with maximal inexact subsets of two different types. We characterize the adjacency relation in terms of maximal inexact subsets. Using this characterization we show that every apartments preserving mapping of $\mathcal{G}$ to $\mathcal{G}'$ is adjacency preserving (two elements of $\mathcal{G}$ are adjacent if and only if their images are adjacent).

Following [Cooperstein, Kasikova and Shult (2005)] we present results of "opposite nature" (characterizations of apartments in terms of the adjacency relation) for certain Grassmannians associated with buildings of classical types.

2.5 Appendix: Gamma spaces

A partial linear space $\Pi = (P, \mathcal{L})$ is called a *gamma space* if it satisfies the following axiom:

(Γ) if a point is collinear with two distinct points of a line then it is collinear with all points of this line.

This axiom implies that for any point $p \in P$ and any line $L \in \mathcal{L}$ the set of all points on L collinear with p is empty, or consists of a single point, or coincides with L. In linear spaces the axiom holds trivially.

Now we establish some elementary properties of gamma spaces. It will be shown later that all Grassmann spaces associated with buildings of classical types are gamma spaces.

Let $\Pi = (P, \mathcal{L})$ be a gamma space. Consider a subset $X \subset P$ consisting of mutually collinear points (a clique in the collinearity graph of Π) and suppose that $|X| \geq 2$. We write $[X]_1$ for the set formed by all points belonging to the lines joining points of X (a point q belongs to $[X]_1$ if there exist distinct points $p, p' \in X$ such that q is on the line pp'). It is clear that $X \subset [X]_1$.

Exercise 2.6. Show that $[X]_1$ is a clique of the collinearity graph.

For every natural $i \geq 2$ we define

$$[X]_i := [[X]_{i-1}]_1.$$

Then $[X]_i$ is contained in $[X]_j$ if $i \leq j$, and, by Exercise 2.6, each $[X]_i$ is a clique of the collinearity graph.

Proposition 2.6. *If X is a clique in the collinearity graph of a certain gamma space and $|X| \geq 2$ then*

$$\langle X \rangle = \bigcup_{i=1}^{\infty} [X]_i$$

is a singular subspace.

Proof. Denote by S the union of all $[X]_i$. If p and q are distinct points of S then $p \in [X]_i$ and $q \in [X]_j$ for certain i, j. It is clear that p and q both belong to $[X]_m$, where $m = \max\{i, j\}$; the line joining these points is contained in $[X]_{m+1}$. Thus S is a singular subspace. Since $\langle X \rangle$ contains every $[X]_i$, we have $X \subset S \subset \langle X \rangle$ which gives the claim. □

Corollary 2.3. *If X is a clique in the collinearity graph of a gamma space $\Pi = (P, \mathcal{L})$ and a point $p \in P$ is collinear with all points of X then p is collinear with all points of $\langle X \rangle$.*

Using Zorn lemma we can show that every graph has maximal cliques and every clique is contained in a certain maximal clique.

Proposition 2.7. *In a gamma space the class of maximal singular subspaces coincides with the class of maximal cliques of the collinearity graph.*

Proof. By Proposition 2.6, every maximal clique X of the collinearity graph is contained in the singular subspace $\langle X \rangle$. Let S be a maximal singular subspace containing $\langle X \rangle$ (Exercise 1.5). Since S is a clique of the collinearity graph, we have $X = \langle X \rangle = S$.

Every maximal singular subspace U is contained in a certain maximal clique Y of the collinearity graph. It was established above that Y is a maximal singular subspace and we get $U = Y$. □

Chapter 3

Classical Grassmannians

It was mentioned in the previous chapter that every thick building of type A_{n-1} ($n \geq 4$) is isomorphic to the flag complex $\Delta(V)$, where V is an n-dimensional vector space over a division ring. The Grassmannians of $\Delta(V)$ are the usual Grassmannians $\mathcal{G}_k(V)$. The associated Grassmann spaces will be denoted by $\mathfrak{G}_k(V)$; their elementary properties will be studied in Section 3.1. Note that $\mathfrak{G}_1(V) = \Pi_V$ and $\mathfrak{G}_{n-1}(V) = \Pi_V^*$.

Let V' be other n-dimensional vector space over a division ring. Following our programme, we will examine collineations of $\mathfrak{G}_k(V)$ to $\mathfrak{G}_k(V')$ and apartments preserving mappings of $\mathcal{G}_k(V)$ to $\mathcal{G}_k(V')$.

The classical Chow's theorem will be proved in Section 3.2. It states that every collineation of $\mathfrak{G}_k(V)$ to $\mathfrak{G}_k(V')$, $1 < k < n-1$, is induced by a semilinear isomorphism of V to V' or a semilinear isomorphism of V to V'^* (the second possibility can be realized only in the case when $n = 2k$). Also we characterize the adjacency relation in terms of the relation to be opposite (as in Theorem 2.4) and show that every bijection of $\mathcal{G}_k(V)$ to $\mathcal{G}_k(V')$, $1 < k < n-1$, preserving the opposite relation is a collineation of $\mathfrak{G}_k(V)$ to $\mathfrak{G}_k(V')$.

If $k = 1, n-1$ then apartments of $\mathcal{G}_k(V)$ are bases of Π_V and Π_V^*, respectively. In Section 3.3 we characterize apartments of $\mathcal{G}_k(V)$, $1 < k < n-1$, in terms of the adjacency relation. This characterization follows from a more general result concerning apartments in *parabolic subspaces* of $\mathfrak{G}_k(V)$. As an application, we describe all subspaces of $\mathfrak{G}_k(V)$ isomorphic to the Grassmann spaces of finite-dimensional vector spaces.

Section 3.4 is dedicated to apartments preserving mappings. All apartments preserving bijections of $\mathcal{G}_k(V)$ to $\mathcal{G}_k(V')$ are collineations of $\mathfrak{G}_k(V)$ to $\mathfrak{G}_k(V')$. If $1 < k < n-1$ then every apartments preserving mapping of $\mathcal{G}_k(V)$ to $\mathcal{G}_k(V')$ is induced by a semilinear embedding of V in V' or a semilinear embedding of V in V'^* (as in Chow's theorem, the second possibility can be realized only in the case when $n = 2k$). There is an example showing that the latter statement fails for $k = 1, n - 1$.

In Sections 3.5–3.7 we demonstrate how our methods work for some constructions rather similar to Grassmannians of finite-dimensional vector spaces — spine spaces, Grassmannians of exchange spaces and the sets of conjugate linear involutions. In particular, Chow's theorem and the theorem on apartments preserving bijections are closely related with classical Dieudonné–Rickart's results on automorphisms of the group $GL(V)$.

In Section 3.8 we extend some results to the case of infinite-dimensional vector spaces focusing on Grassmannians formed by linear subspaces with infinite dimension and codimension.

3.1   Elementary properties of Grassmann spaces

Let V be an n-dimensional vector space over a division ring and $3 \le n < \infty$. Let also $k \in \{1, \ldots, n-1\}$. Recall that two k-dimensional linear subspaces of V (elements of the Grassmannian $\mathcal{G}_k(V)$) are *adjacent* if their intersection is $(k-1)$-dimensional; the latter is equivalent to the fact that the sum of these linear subspaces is $(k+1)$-dimensional.

Now suppose that M and N are incident linear subspaces of V and

$$\dim M < k < \dim N$$

(possible $M = 0$ or $N = V$). We define

$$[M, N]_k := \{ \, S \in \mathcal{G}_k(V) \; : \; M \subset S \subset N \, \}.$$

If $M = 0$ or $N = V$ then instead of $[M, N]_k$ we will write

$$\langle N]_k \quad \text{or} \quad [M \rangle_k,$$

respectively. In the case when

$$\dim M = k - 1 \quad \text{and} \quad \dim N = k + 1,$$

the set $[M, N]_k$ is a *line* of $\mathcal{G}_k(V)$. This line contains at least three points: we choose two linearly independent vectors $x, y \in N$ such that N is spanned by M and x, y then

$$\langle M, x \rangle, \; \langle M, y \rangle, \; \langle M, x + y \rangle$$

are three distinct points on the line.

The set of all such lines will be denoted by $\mathcal{L}_k(V)$. Two distinct elements of $\mathcal{G}_k(V)$ are collinear (joined by a line) if and only if they are adjacent; and for any adjacent $S, U \in \mathcal{G}_k(V)$ there is the unique line

$$[S \cap U, S + U]_k$$

containing them. The pair

$$\mathfrak{G}_k(V) := (\mathcal{G}_k(V), \mathcal{L}_k(V))$$

is a partial linear space. It is clear that $\mathfrak{G}_1(V) = \Pi_V$ and $\mathfrak{G}_{n-1}(V) = \Pi_V^*$; moreover,

$$\mathfrak{G}_k(V^*) \quad \text{and} \quad \mathfrak{G}_{n-k}(V)$$

are canonically isomorphic (by the annihilator mapping).

The partial linear spaces $\mathfrak{G}_k(V)$, $k \in \{1, \ldots, n-1\}$, are the Grassmann spaces of the building $\Delta(V)$. The associated Grassmann graphs (the collinearity graphs of $\mathfrak{G}_k(V)$) will be denoted by $\Gamma_k(V)$.

Recall that three distinct mutually collinear points of a partial linear space form a *triangle* if they are not collinear (in other words, these points span a plane).

Lemma 3.1. *Let $1 < k < n-1$. For any triangle S_1, S_2, S_3 in $\mathfrak{G}_k(V)$ only one of the following two possibilities is realized:*

(1) *a star-triangle: there is a $(k-1)$-dimensional linear subspace contained in each S_i,*

(2) *a top-triangle: there is a $(k+1)$-dimensional linear subspace containing all S_i.*

Proof. The fulfillment of both (1) and (2) implies that S_1, S_2, S_3 are collinear. Therefore, one of these conditions does not hold. Suppose that S_3 is not contained in the $(k+1)$-dimensional linear subspace $S_1 + S_2$. Then the dimension of the linear subspace

$$U := (S_1 + S_2) \cap S_3$$

is not greater than $k - 1$. Since S_3 is adjacent with S_1 and S_2, U is a $(k-1)$-dimensional linear subspace contained in S_1 and S_2. $\square$

By Proposition 2.1, the Grassmann space $\mathfrak{G}_k(V)$ is connected. Now we prove the following.

Proposition 3.1. *The Grassmann space $\mathfrak{G}_k(V)$ is a connected gamma space. For any k-dimensional linear subspaces $S, U \subset V$ the distance between S and U is equal to*

$$\dim(S + U) - k = k - \dim(S \cap U).$$

Proof. The case $k = 1, n - 1$ is trivial. Let $1 < k < n - 1$. If $S \in \mathcal{G}_k(V)$ is adjacent with two distinct points S_1, S_2 of a certain line $[M, N]_k$ and S does not belong to this line then

$$M = S_1 \cap S_2 \subset S \quad \text{or} \quad S \subset S_1 + S_2 = N$$

$(S, S_1, S_2$ form a star-triangle or a top-triangle, respectively). In each of these cases, S is adjacent with all points of the line $[M, N]_k$. Thus the axiom (Γ) holds (Section 2.5).

For any two linear subspaces $S, U \subset V$ there is a base $B \subset V$ such that S and U are spanned by subsets of B (Proposition 1.4). Suppose that the linear subspaces S and U both are k-dimensional. Let

$$X \cup \{x_1, \dots, x_{2m}\}$$

be a subset of B such that

$$S \cap U = \langle X \rangle,$$
$$S = \langle X, x_1, \dots, x_m \rangle,$$
$$U = \langle X, x_{m+1}, \dots, x_{2m} \rangle.$$

The k-dimensional linear subspaces

$$S_i := \langle X, x_{1+i}, \dots, x_{m+i} \rangle, \quad i = 0, \dots, m,$$

form a path in $\Gamma_k(V)$ connecting S and U. For every path

$$S = U_0, U_1, \dots, U_l = U$$

we have

$$\dim(U_0 \cap \cdots \cap U_l) \geq k - l.$$

On the other hand, $U_0 \cap \cdots \cap U_l$ is contained in the linear subspace $S \cap U$ whose dimension is equal to $k - m$. Thus $k - l \leq k - m$ and $m \leq l$ which means that $d(S, U) = m$. □

Remark 3.1 (Grassmann embedding). Suppose that V is a vector space over a field and consider the exterior power vector space $\wedge^k V$. If vectors $x_1, \dots, x_k$ and $y_1, \dots, y_k$ span the same k-dimensional linear subspace of V then

$$x_1 \wedge \cdots \wedge x_k = a(y_1 \wedge \cdots \wedge y_k)$$

for a certain non-zero scalar a. This implies that the *Grasmann mapping*

$$g_k : \mathcal{G}_k(V) \to \mathcal{G}_1(\wedge^k V)$$
$$\langle x_1, \dots, x_k \rangle \to \langle x_1 \wedge \cdots \wedge x_k \rangle$$

is well-defined. This mapping is injective and the image of every line of $\mathfrak{G}_k(V)$ is a line of $\Pi_{\wedge^k V}$. Thus g_k is an embedding of the Grassmann space $\mathfrak{G}_k(V)$ in the projective space $\Pi_{\wedge^k V}$.

Now we describe maximal singular subspaces of $\mathfrak{G}_k(V)$, $1 < k < n - 1$. Since $\mathfrak{G}_k(V)$ is a gamma space, the class of maximal singular subspaces of $\mathfrak{G}_k(V)$ coincides with the class of maximal cliques of the Grassmann graph $\Gamma_k(V)$ (Proposition 2.7).

Example 3.1. For every $(k+1)$-dimensional linear subspace $N \subset V$ the set $\langle N]_k$ will be called a *top*. This is a singular subspace of $\mathfrak{G}_k(V)$ isomorphic to a k-dimensional projective space. If $S \in \mathcal{G}_k(V)$ does not belong to $\langle N]_k$ then

$$\dim(S \cap N) \leq k - 1$$

and there exist elements of $\langle N]_k$ which are not adjacent with S. This means that $\langle N]_k$ is a maximal singular subspace of $\mathfrak{G}_k(V)$. Every triangle of $\langle N]_k$ is a top-triangle.

Example 3.2. For every $(k-1)$-dimensional linear subspace $M \subset V$ the set $[M\rangle_k$ is said to be a *star*. This is a singular subspace of $\mathfrak{G}_k(V)$ isomorphic to an $(n - k)$-dimensional projective space. Stars are maximal singular subspaces, since the canonical collineation between $\mathfrak{G}_k(V)$ and $\mathfrak{G}_{n-k}(V^*)$ (the annihilator mapping) sends stars to tops and tops to stars. Every triangle of a star is a star-triangle.

Proposition 3.2 ([Chow (1949)]). *Every maximal singular subspace of* $\mathfrak{G}_k(V)$, $1 < k < n - 1$, *is a star or a top.*

Proof. It was noted above that the class of maximal singular subspaces coincides with the class of maximal cliques of the Grassmann graph. Hence it is sufficient to show that every clique of the Grassmann graph is contained in a star or a top.

Let $\mathcal{X}$ be a clique of $\Gamma_k(V)$ and S_1, S_2 be distinct elements of $\mathcal{X}$. Suppose that $\mathcal{X}$ is not contained in a star. In this case, there exists $S_3 \in \mathcal{X}$ which does not contain $S_1 \cap S_2$ and S_1, S_2, S_3 form a top-triangle. Then S_3 is contained in the $(k + 1)$-dimensional linear subspace

$$N := S_1 + S_2.$$

If a k-dimensional linear subspace S does not belong to the top $\langle N]_k$ then

$$\dim(N \cap S) \leq k - 1$$

and S is not adjacent with at least one of S_i. Thus $S \notin \mathcal{X}$ and our clique is a subset of $\langle N]_k$. $\qquad\square$

Example 3.3. Let M and N be incident linear subspaces of V satisfying

$$\dim M < k < \dim N$$

(possible $M = 0$ or $N = V$). Then $[M, N]_k$ is a subspace of $\mathfrak{G}_k(V)$; subspaces of such type are called *parabolic* [Cooperstein, Kasikova and Shult (2005)]. The parabolic subspace $[M, N]_k$ is isomorphic to the Grassmann space $\mathfrak{G}_{k-m}(N/M)$, where m is the dimension of M. This subspace is singular only in the case when M is $(k-1)$-dimensional or N is $(k+1)$-dimensional. By Proposition 3.2, every singular subspace of $\mathfrak{G}_k(V)$ is parabolic. The canonical collineation between $\mathfrak{G}_k(V)$ and $\mathfrak{G}_{n-k}(V^*)$ transfers parabolic subspaces to parabolic subspaces.

The following result [Cooperstein, Kasikova and Shult (2005); Żynel (2000)] generalizes Proposition 3.2. It will be proved in Section 3.3.

Theorem 3.1. *Every subspace of $\mathfrak{G}_k(V)$, $1 < k < n - 1$, isomorphic to the Grassmann space of a finite-dimensional vector space is parabolic.*

Let S and U be adjacent elements of $\mathcal{G}_k(V)$, $1 < k < n - 1$. The set of all elements of $\mathcal{G}_k(V)$ adjacent with both S, U is the union of all maximal cliques of the Grassmann graph (all maximal singular subspaces of $\mathfrak{G}_k(V)$) containing S and U. By Proposition 3.2, this is

$$[S \cap U\rangle_k \cup \langle S + U]_k.$$

A k-dimensional linear subspace of V is adjacent with all elements of this set if and only if it belongs to the line

$$[S \cap U, S + U]_k$$

joining S and U.

Denote by $\mathcal{X}^\sim$ the set consisting of all elements of $\mathcal{G}_k(V)$ adjacent with each element of a subset $\mathcal{X} \subset \mathcal{G}_k(V)$. We get the following characterization of lines in terms of the adjacency relation.

Proposition 3.3. *For any adjacent $S, U \in \mathcal{G}_k(V)$, $1 < k < n - 1$, the line joining S and U coincides with the subset $\{S, U\}^{\sim\sim}$.*

The intersection of two distinct stars $[M\rangle_k$ and $[M'\rangle_k$ contains at most one element; this intersection is not empty if and only if M, M' are adjacent elements of $\mathcal{G}_{k-1}(V)$. The same holds for the intersection of two distinct tops. The intersection of the star $[M\rangle_k$ and the top $\langle N]_k$ is empty or a line; the second possibility is realized only in the case when $M \subset N$.

3.2 Collineations of Grassmann spaces

Throughout the section we suppose that V and V' are n-dimensional vector spaces over division rings and $4 \le n < \infty$. We will study collineations between the Grassmann spaces $\mathfrak{G}_k(V)$ and $\mathfrak{G}_k(V')$ for $1 < k < n - 1$.

3.2.1 *Chow's theorem*

For every semilinear isomorphism $u : V \to V'$ the bijection

$$(u)_k : \mathcal{G}_k(V) \to \mathcal{G}_k(V')$$

(Subsection 1.3.2) is a collineation of $\mathfrak{G}_k(V)$ to $\mathfrak{G}_k(V')$. By duality, it can be considered as the collineation of $\mathfrak{G}_{n-k}(V^*)$ to $\mathfrak{G}_{n-k}(V'^*)$ induced by the contragradient $\check{u}$ (Subsection 1.3.3).

Let $l : V \to V'^*$ be a semilinear isomorphism. By duality, we can identify $(l)_k$ with the collineation of $\mathfrak{G}_k(V)$ to $\mathfrak{G}_{n-k}(V')$ sending each S to $l(S)^0$. This is a collineation of $\mathfrak{G}_k(V)$ to $\mathfrak{G}_k(V')$ if $n = 2k$.

Theorem 3.2 ([Chow (1949)]). *Let* $1 < k < n - 1$. *Then every isomorphism of* $\Gamma_k(V)$ *to* $\Gamma_k(V')$ *is the collineation of* $\mathfrak{G}_k(V)$ *to* $\mathfrak{G}_k(V')$ *induced by a semilinear isomorphism of* V *to* V' *or a semilinear isomorphism of* V *to* V'^*; *the second possibility can be realized only in the case when* $n = 2k$.

Remark 3.2. In the cases when $k = 1, n - 1$, every bijection of $\mathcal{G}_k(V)$ to $\mathcal{G}_k(V')$ is an isomorphism of $\Gamma_k(V)$ to $\Gamma_k(V')$.

It follows immediately from Theorem 3.2 that every collineation of $\mathfrak{G}_k(V)$ to $\mathfrak{G}_k(V')$ is induced by a semilinear isomorphism of V to V' or V'^*. This statement generalizes the classical version of the Fundamental Theorem of Projective Geometry (Corollary 1.2).

Proof. Let f be an isomorphism of $\Gamma_k(V)$ to $\Gamma_k(V')$. By Proposition 3.3, f is a collineation of $\mathfrak{G}_k(V)$ to $\mathfrak{G}_k(V')$. Then f and f^{-1} map maximal singular subspaces to maximal singular subspaces. Recall that maximal singular subspaces of our Grassmann spaces are stars and tops. The intersection of two distinct maximal singular subspaces is empty, or a single point, or a line; the third possibility is realized only in the case when these singular subspaces are of different types (one of them is a star and the other is a top) and the associated $(k - 1)$-dimensional and $(k + 1)$-dimensional linear subspaces are incident.

Now suppose that for a $(k-1)$-dimensional linear subspace $S \subset V$ the image of the star $[S\rangle_k$ is a star. Let U be a $(k-1)$-dimensional linear subspace adjacent with S. We choose a $(k+1)$-dimensional linear subspace N containing S and U. The stars $[S\rangle_k$ and $[U\rangle_k$ intersect the top $\langle N]_k$ by lines. Since $[S\rangle_k$ goes to a star, the image of $\langle N]_k$ is a top; hence $f([U\rangle_k)$ is a star. By connectedness (Proposition 3.1), the same holds for every $(k-1)$-dimensional linear subspace $U \subset V$. Similarly, we establish that all stars go to tops if the image of a certain star of $\mathfrak{G}_k(V)$ is a top. Therefore, one of the following possibilities is realized:

(A) stars go to stars,
(B) stars go to tops.

The same arguments show that tops go to tops and tops go to stars in the cases (A) and (B), respectively.

Case (A). In this case, there exists a bijection

$$f_{k-1} : \mathcal{G}_{k-1}(V) \to \mathcal{G}_{k-1}(V')$$

such that

$$f([S\rangle_k) = [f_{k-1}(S)\rangle_k$$

for all $S \in \mathcal{G}_{k-1}(V)$. For every $U \in \mathcal{G}_k(V)$

$$S \in \langle U]_{k-1} \Leftrightarrow U \in [S\rangle_k \Leftrightarrow f(U) \in [f_{k-1}(S)\rangle_k \Leftrightarrow f_{k-1}(S) \in \langle f(U)]_{k-1}$$

and we have

$$f_{k-1}(\langle U]_{k-1}) = \langle f(U)]_{k-1}.$$

Thus f_{k-1} and the inverse mapping send tops to tops which implies that f_{k-1} is an isomorphism of type (A) between the Grassmann graphs $\Gamma_{k-1}(V)$ and $\Gamma_{k-1}(V')$. Step by step, we get a sequence of such isomorphisms

$$f_i : \mathcal{G}_i(V) \to \mathcal{G}_i(V'), \quad i = k, \ldots, 1,$$

where $f_k = f$ and

$$f_i([S\rangle_i) = [f_{i-1}(S)\rangle_i$$

for all $S \in \mathcal{G}_{i-1}(V)$ if $i > 1$. Then

$$f_{i-1}(\langle U]_{i-1}) = \langle f_i(U)]_{i-1} \qquad (3.1)$$

for each $U \in \mathcal{G}_i(V)$; in particular, f_1 is a collineation of Π_V to $\Pi_{V'}$. By the Fundamental Theorem of Projective Geometry, f_1 is induced by a semilinear isomorphism $l : V \to V'$. We can prove that $f_i = (l)_i$ induction by i. Indeed, if f_{i-1} is induced by l then for every $U \in \mathcal{G}_i(V)$ we have

$$f_{i-1}(\langle U]_{i-1}) = \langle l(U)]_{i-1}$$

and (3.1) gives the claim.

Case (B). Stars and tops are projective spaces of dimension $n - k$ and k, respectively. Since f is a collineation and preserves the dimensions of singular subspaces, we have $n = 2k$. By duality, f can be considered as a colleniation of $\mathfrak{G}_k(V)$ to $\mathfrak{G}_k(V'^*)$ sending stars to stars. Hence it is induced by a semilinear isomorphism of V to V'^*. $\square$

The following remarkable result is closely related with Chow's theorem.

Theorem 3.3 ([Huang (1998)]). *Let* $1 < k < n - 1$. *If a surjection of* $\mathcal{G}_k(V)$ *to* $\mathcal{G}_k(V')$ *sends adjacent elements to adjacent elements then it is a collineation of* $\mathfrak{G}_k(V)$ *to* $\mathfrak{G}_k(V')$; *in particular, every semicollineation of* $\mathfrak{G}_k(V)$ *to* $\mathfrak{G}_k(V')$ *is a collineation.*

3.2.2 *Chow's theorem for linear spaces*

There is an analogue of Chow's theorem for linear spaces. Let $\Pi = (P, \mathcal{L})$ be a linear space. Two distinct lines of Π are non-intersecting or have a common point; in the second case, these lines are said to be *adjacent*. Every pair of adjacent lines spans a plane. However, in contrast to the projective case, there are planes containing pairs of non-adjacent lines (for example, parallel lines in affine planes). The *Grassmann graph* $\Gamma_1(\Pi)$ is the graph whose vertex set is $\mathcal{L}$ and whose edges are pairs of adjacent lines. It is not difficult to prove that $\Gamma_1(\Pi)$ is connected and every maximal clique of this graph is a star or a subset of a top (a star is formed by all lines passing through a point, a top is the set of all lines contained in a plane). Every collineation between linear spaces induces an isomorphism between the associated Grassmann graphs; this isomorphism sends stars to stars.

Theorem 3.4 ([Havlicek (1999)]). *Let* $\Pi = (P, \mathcal{L})$ *and* $\Pi' = (P', \mathcal{L}')$ *be linear spaces of the same finite dimension* $n \geq 3$. *Suppose that* $f : \mathcal{L} \to \mathcal{L}$ *is an isomorphism of* $\Gamma_1(\Pi)$ *to* $\Gamma_1(\Pi')$. *Then one of the following possibilities is realized:*

- *The isomorphism* f *maps stars to stars and it is induced by a collineation of* Π *to* Π'.
- *All tops are maximal cliques and* f *transfers stars to tops and tops to stars. In this case,* Π *and* Π' *both are 3-dimensional generalized projective spaces and* f *is induced by a collineation of* Π *to the generalized projective space dual to* Π'.

Remark 3.3. A *generalized projective space* can be defined as a linear space where any two lines contained in a plane have a common point (if every line of a generalized projective space contains at least three points then it is a projective space). Every finite-dimensional generalized projective space is the union of a finite collection of the following "components": points, lines containing more than two points, projective spaces; points from distinct components are joined by lines of cardinality 2 [Buekenhout and Cameron (1995)]. Suppose that $\Pi = (P, \mathcal{L})$ is a generalized projective space of dimension 3 and denote by P^* the set of all planes of Π. A subset of P^* is called a *line* if it consists of all planes containing a certain line $L \in \mathcal{L}$; the set of all such lines is denoted by $\mathcal{L}^*$. Then $\Pi^* := (P^*, \mathcal{L}^*)$ is a generalized projective space of dimension 3; this generalized projective space is called *dual* to Π. There is natural one-to-one correspondence between elements of $\mathcal{L}$ and $\mathcal{L}^*$. Since two lines of Π are adjacent if and only if the corresponding lines of Π^* are adjacent, $\mathcal{L}^*$ can be identified with $\mathcal{L}$ and every collineation of Π to Π^* induces an automorphism of the Grassmann graph $\Gamma_1(\Pi)$.

3.2.3   *Applications of Chow's theorem*

Proposition 3.4. *Let* $k, m \in \{1, \ldots, n-1\}$ *and* $k \neq m$. *Let also*

$$g_k : \mathcal{G}_k(V) \to \mathcal{G}_k(V') \quad and \quad g_m : \mathcal{G}_m(V) \to \mathcal{G}_m(V')$$

be bijections satisfying the following condition: $S \in \mathcal{G}_k(V)$ *and* $U \in \mathcal{G}_m(V)$ *are incident if and only if* $g_k(S)$ *and* $g_m(U)$ *are incident. Then there exists a semilinear isomorphism* $l : V \to V'$ *such that*

$$g_k = (l)_k \quad and \quad g_m = (l)_m,$$

in other words, g_k *and* g_m *are induced by the same semilinear isomorphism of* V *to* V'.

Proof.   By symmetry, we can restrict ourselves to the case when $k < m$. It is clear that

$$g_k(\langle U]_k) = \langle g_m(U)]_k$$

for all $U \in \mathcal{G}_m(V)$. Let us define

$$\mathcal{G}_{k,m}(V) := \bigcup_{i=k}^{m} \mathcal{G}_i(V) \quad \text{and} \quad \mathcal{G}_{k,m}(V') := \bigcup_{i=k}^{m} \mathcal{G}_i(V').$$

For every $U \in \mathcal{G}_{k,m}(V)$ there exists $g(U) \in \mathcal{G}_{k,m}(V')$ such that

$$g_k(\langle U]_k) = \langle g(U)]_k.$$

Indeed, we choose m-dimensional linear subspaces $U_1, \ldots, U_i$ satisfying
$$U = U_1 \cap \cdots \cap U_i,$$
then
$$g(U) := g_m(U_1) \cap \cdots \cap g_m(U_i)$$
is as required. The mapping
$$g : \mathcal{G}_{k,m}(V) \to \mathcal{G}_{k,m}(V')$$
is bijective and its restrictions to $\mathcal{G}_k(V)$ and $\mathcal{G}_m(V)$ coincide with g_k and g_m, respectively. For all $M, N \in \mathcal{G}_{k,m}(V)$ we have
$$M \subset N \iff g(M) \subset g(N);$$
in particular, g transfers $\mathcal{G}_{k+1}(V)$ to $\mathcal{G}_{k+1}(V')$. This means that g_k is a collineation of Π_V to $\Pi_{V'}$ if $k = 1$. In the case when $k > 1$, the mapping g_k is an isomorphism of $\Gamma_k(V)$ to $\Gamma_k(V')$ (since g_k and the inverse mapping send tops to tops). Therefore, g_k is induced by a semilinear isomorphism $l : V \to V'$; an easy verification shows that $g_m = (l)_m$. $\qquad\square$

By duality, we have the following.

Corollary 3.1. *Let $k, m \in \{1, \ldots, n-1\}$ and $k \neq m$. Let also*
$$g_k : \mathcal{G}_k(V) \to \mathcal{G}_{n-k}(V') \quad and \quad g_m : \mathcal{G}_m(V) \to \mathcal{G}_{n-m}(V')$$
be bijections such that $S \in \mathcal{G}_k(V)$ and $U \in \mathcal{G}_m(V)$ are incident if and only if $g_k(S)$ and $g_m(U)$ are incident. Then g_k and g_m are induced by the same semilinear isomorphism of V to V'^*.

Let Ω be a non-degenerated reflexive form defined on V and $\perp$ be the associated orthogonal relation (Subsection 1.5.2). Consider the bijective transformation of $\mathcal{G}(V)$ sending each linear subspace $S \subset V$ to $S^\perp$. Suppose that f is the restriction of this transformation to $\mathcal{G}_k(V)$. Then f is a collineation of $\mathfrak{G}_k(V)$ to $\mathfrak{G}_{n-k}(V)$ satisfying the following condition:

(P) for any $S, U \in \mathcal{G}_k(V)$ the linear subspaces $S, f(U)$ are incident if and only if $U, f(S)$ are incident.

If $k = 1$ then f is a polarity (Subsection 1.5.3). The following result generalizes Proposition 1.13.

Theorem 3.5 ([Pankov 1 (2004)]). *Let $n \neq 2k$ and f be a bijection of $\mathcal{G}_k(V)$ to $\mathcal{G}_{n-k}(V)$ satisfying the condition (P). Then there exists a non-degenerate reflexive form Ω such that the restriction of the transformation $S \to S^\perp$ (where $\perp$ is the orthogonal relation associated with Ω) to $\mathcal{G}_k(V)$ coincides with f.*

Proof.   Since $k \neq n - k$, we can apply Corollary 3.1 to the bijections

$$f : \mathcal{G}_k(V) \to \mathcal{G}_{n-k}(V) \ \text{ and } \ f^{-1} : \mathcal{G}_{n-k}(V) \to \mathcal{G}_k(V).$$

So, f and f^{-1} both are induced by a semilinear isomorphism $u : V \to V^*$. Let us consider the bijective transformation h of $\mathcal{G}(V)$ which sends every linear subspace S to $u(S)^0$. The restrictions of h to $\mathcal{G}_k(V)$ and $\mathcal{G}_{n-k}(V)$ coincide with f and f^{-1}, respectively. Therefore, the restriction of h^2 to $\mathcal{G}_k(V)$ is identity. This implies that h^2 is identity (we leave the details for the reader). The latter means that the sesquilinear form defined by u is reflexive. This reflexive form is as required. $\qquad \square$

Remark 3.4. In the case when $n = 2k$, the condition (P) is equivalent to the fact that f is an involution of $\mathcal{G}_k(V)$ (f^2 is identity) and the statement given above fails.

3.2.4   *Opposite relation*

Let $1 < k < n - 1$. Two vertices of the Grassmann graph $\Gamma_k(V)$ (elements of the Grassmannian $\mathcal{G}_k(V)$) are said to be *opposite* if the distance between them is maximal (equal to the diameter of the graph). By Proposition 3.1, the diameter of $\Gamma_k(V)$ is equal to

$$\begin{cases} k & \text{if } \ 2k \leq n \\ n - k & \text{if } \ 2k > n. \end{cases}$$

Therefore, in the case when $2k \leq n$, two elements of $\mathcal{G}_k(V)$ are opposite if and only if their intersection is zero.

The adjacency relation can be characterized in terms of the relation to be opposite.

Theorem 3.6. *Let $1 < k < n - 1$. For any distinct k-dimensional linear subspaces $S_1, S_2 \subset V$ the following conditions are equivalent:*

(1) *S_1 and S_2 are adjacent,*
(2) *there exists $S \in \mathcal{G}_k(V) \setminus \{S_1, S_2\}$ such that every $U \in \mathcal{G}_k(V)$ opposite to S is opposite to at least one of S_i.*

Proof.   Since the annihilator mapping of $\mathcal{G}_k(V)$ to $\mathcal{G}_{n-k}(V^*)$ is an isomorphism between the associated Grassmann graphs, we can restrict ourselves to the case when $2k \leq n$. In this case, two k-dimensional linear subspaces are opposite if and only if their intersection is zero.

(1) $\implies$ (2). If S_1 and S_2 are adjacent elements of $\mathcal{G}_k(\Pi)$ then every $S \in \mathcal{G}_k(V) \setminus \{S_1, S_2\}$ belonging to the line joining S_1 and S_2 is as required. Indeed, if $U \in \mathcal{G}_k(V)$ is opposite to S then it intersects $S_1 + S_2$ in a certain 1-dimensional linear subspace P; since $S_1 \cap S_2 \subset S$, this linear subspace is not contained $S_1 \cap S_2$ and

$$(S_1 \cap S_2) + P$$

is a unique point of the line joining S_1 and S_2 which is not opposite to U.

(2) $\implies$ (1). The proof of this implication will be given in several steps.

First we establish that for any 1-dimensional linear subspaces $P_1 \subset S_1$ and $P_2 \subset S_2$ the sum $P_1 + P_2$ has a non-zero intersection with S.

Suppose that this intersection is zero. Then $P_1 + P_2$ is contained in a k-dimensional linear subspace opposite to S. By our hypothesis, this linear subspace is opposite to at least one of S_i. Thus there is P_i which is not contained in S_i, a contradiction.

In particular, every 1-dimensional linear subspace $P \subset S_1 \cap S_2$ is contained in S (we take $P_1 = P_2 = P$); hence

$$S_1 \cap S_2 \subset S. \tag{3.2}$$

Our second step is to show that

$$\dim(S \cap S_1) = \dim(S \cap S_2) = k - 1. \tag{3.3}$$

It is sufficient to establish that every 2-dimensional linear subspace contained in S_i $(i = 1, 2)$ has a non-zero intersection with S.

Let us take a 2-dimensional linear subspace $U \subset S_1$ and a 1-dimensional linear subspace $P_2 \subset S_2$ which is not contained in S. By (3.2), P_2 is not contained in S_1; hence $U + P_2$ is 3-dimensional. Let P_1 and Q_1 be distinct 1-dimensional linear subspaces of U. By the first step of the proof, the linear subspaces $P_1 + P_2$ and $Q_1 + P_2$ meet S in 1-dimensional linear subspaces P and Q, respectively. It is clear that $P \neq Q$. Since U and $P + Q$ are 2-dimensional linear subspaces contained in the 3-dimensional linear subspace $U + P_2$, their intersection is non-zero. The inclusion $P + Q \subset S$ gives the claim.

Now we choose 1-dimensional linear subspaces $P_1 \subset S_1$ and $P_2 \subset S_2$ such that

$$S_i = (S \cap S_i) + P_i$$

(this is possible by (3.3)). Then $P_1 + P_2$ has a non-zero intersection with S (by the first step of the proof) and $S + P_1$ contains P_2. Thus S_1 and S_2 both are contained in the $(k+1)$-dimensional linear subspace $S + P_1$ which means that they are adjacent. $\qquad\square$

Corollary 3.2. *Let* $1 < k < n - 1$*. Every bijection of* $\mathcal{G}_k(V)$ *to* $\mathcal{G}_k(V')$ *preserving the relation to be opposite (two elements of* $\mathcal{G}_k(V)$ *are opposite if and only if their images are opposite) is a collineation of* $\mathfrak{G}_k(V)$ *to* $\mathfrak{G}_k(V')$*.*

Remark 3.5. In the case when $k = 2, n - 2$, two distinct elements of $\mathcal{G}_k(V)$ are adjacent or opposite and the latter statement is trivial.

Theorem 3.6 can be reformulated in the following form.

Theorem 3.7. *Let* $1 < k < n - 1$*. For any distinct* k*-dimensional linear subspaces* $S_1, S_2 \subset V$ *the following conditions are equivalent:*

(1) S_1 *and* S_2 *are adjacent,*
(3) *there exists* $S \in \mathcal{G}_k(V) \setminus \{S_1, S_2\}$ *such that every complement of* S *is a complement to at least one of* S_i*.*

Proof. As in the proof of Theorem 3.6, we can restrict ourselves to the case when $2k \le n$. The implication (1) $\Longrightarrow$ (3) is obvious. Since an element of $\mathcal{G}_k(V)$ is opposite to $S \in \mathcal{G}_k(V)$ if and only if it is contained in a complement of S (this is true only in the case when $2k \le n$), we have (3) $\Longrightarrow$ (2) and Theorem 3.6 gives the claim. $\square$

Remark 3.6. Theorem 3.6 was first proved in [Blunck and Havlicek (2005)] for the case when $n = 2k$, and it was shown later [Havlicek and Pankov (2005)] that Blunck–Havlicek's method works for the general case. In [Huang and Havlicek (2008)] the same characterization was obtained for an abstract graph satisfying certain technical conditions; an easy verification shows that these conditions hold for the Grassmann graph $\Gamma_k(V)$.

Remark 3.7. Let m be a positive integer which is less than the diameter of $\Gamma_k(V)$. For a subset $\mathcal{X} \subset \mathcal{G}_k(V)$ we define

$$\mathcal{X}^m := \{ \, Y \in \mathcal{G}_k(V) \, : \, d(X, Y) \le m \quad \forall \, X \in \mathcal{X} \, \}.$$

By [Lim (2010)], for any $S, U \in \mathcal{G}_k(V)$ satisfying $0 < d(S, U) \le m$ one of the following possibilities is realized: S, U are adjacent and $(\{S, U\}^m)^m$ is the line joining S with U, or they are non-adjacent and $(\{S, U\}^m)^m$ coincides with $\{S, U\}$. As a consequence, we have the following result [Lim (2010)]: if $f : \mathcal{G}_k(V) \to \mathcal{G}_k(V')$ is a surjection such that

$$d(S, U) \le m \iff d(f(S), f(U)) \le m$$

for any $S, U \in \mathcal{G}_k(V)$ then f is an isomorphism of $\Gamma_k(V)$ to $\Gamma_k(V')$.

By the remarks given above, Theorem 3.6 is a partial case of more general results. On the other hand, our proof of Theorem 3.6 is short and it can be modified for Grassmannians of infinite-dimensional vector spaces, see Section 3.8. The same idea also will be exploited in Section 4.7. By these reasons, we present this proof here.

3.3 Apartments

3.3.1 *Basic properties*

Let V be an n-dimensional vector space over a division ring and $3 \le n < \infty$. Let also $B = \{x_1, \ldots, x_n\}$ be a base of V. Denote by $\mathcal{A}_k$ the associated *apartment* of the Grassmannian $\mathcal{G}_k(V)$, it consists all k-dimensional linear subspaces $\langle x_{i_1}, \ldots, x_{i_k} \rangle$ and

$$|\mathcal{A}_k| = \binom{n}{k}.$$

Remark 3.8. The apartment of $\mathcal{G}_k(V)$ defined by a base $B' \subset V$ coincides with $\mathcal{A}_k$ if and only if the vectors of B' are scalar multiples of the vectors of B. Thus there is a one-to-one correspondence between apartments and bases of the projective space Π_V.

Exercise 3.1. Show that for any $S, U \in \mathcal{G}_k(V)$ the intersection of all apartments of $\mathcal{G}_k(V)$ containing S and U coincides with $\{S, U\}$.

The annihilator mapping transfers $\mathcal{A}_k$ to the apartment of $\mathcal{G}_{n-k}(V^*)$ associated with the dual base B^* (Subsection 1.1.3).

The restriction of the Grassmann graph $\Gamma_k(V)$ to a subset $\mathcal{X} \subset \mathcal{G}_k(V)$ will be denoted by $\Gamma(\mathcal{X})$. It is clear that $\Gamma(\mathcal{A}_k)$ is isomorphic to the *Johnson graph* $J(n, k)$ (the graph whose vertex set is formed by all k-element subsets of $\{1, \ldots, n\}$ and two such subsets are connected by an edge if their intersection consists of $k - 1$ elements). If $k = 1, n - 1$ then $\mathcal{A}_k$ is a base of Π_V or Π_V^*, respectively. If $1 < k < n - 1$ then the maximal cliques of $\Gamma(\mathcal{A}_k)$ are the intersections of $\mathcal{A}_k$ with the stars

$$[M\rangle_k, \quad M \in \mathcal{A}_{k-1}$$

and the tops

$$\langle N]_k, \quad N \in \mathcal{A}_{k+1};$$

they will be called *stars* and *tops* of $\mathcal{A}_k$ (respectively). Every maximal clique of $\Gamma(\mathcal{A}_k)$ is a base of the corresponding maximal singular subspace of $\mathfrak{G}_k(V)$.

Proposition 3.5. *The Grassmann space $\mathfrak{G}_k(V)$ is spanned by every apartment of $\mathcal{G}_k(V)$.*

Proof. The case $k = 1$ is trivial and we prove the statement induction by k. Let $B = \{x_1, \ldots, x_n\}$ be a base of V and $\mathcal{A}$ be the associated apartment of $\mathcal{G}_k(V)$. Denote by $\mathcal{X}$ the subspace of $\mathfrak{G}_k(V)$ spanned by $\mathcal{A}$ and define $P_i := \langle x_i \rangle$ for every i. Since $[P_i\rangle_k$ can be identified with the Grassmann space $\mathfrak{G}_{k-1}(V/P_i)$, the inductive hypothesis implies that $[P_i\rangle_k$ is spanned by $\mathcal{A} \cap [P_i\rangle_k$. Therefore,

$$[P_i\rangle_k \subset \mathcal{X}$$

for every i. Now consider a k-dimensional linear subspace $S \subset V$ such that $x_i \notin S$ for all i. We take any $(k-1)$-dimensional linear subspace $M \subset S$ and define

$$S_i := M + P_i$$

for every i. Then $S_1, \ldots, S_n$ span the star $[M\rangle_k$ (otherwise, $S_1 + \cdots + S_n$ is a proper subspace of V which contradicts the fact that B is a base). Every S_i belongs to $\mathcal{X}$ and we get $S \in [M\rangle_k \subset \mathcal{X}$ which completes our proof. □

Now we investigate apartments of parabolic subspaces (Example 3.3). Let M and N be incident linear subspaces of V such that

$$\dim M = m < k < l = \dim N.$$

We take any base of V whose subsets span M and N; the intersection of the associated apartment of $\mathcal{G}_k(V)$ with the parabolic subspace $[M, N]_k$ is called an *apartment* of $[M, N]_k$. The natural collineation of $[M, N]_k$ to the Grassmann space $\mathfrak{G}_{k-m}(N/M)$ establishes a one-to-one correspondence between apartments of $[M, N]_k$ and apartments of $\mathcal{G}_{k-m}(N/M)$. Proposition 3.5 guarantees that $[M, N]_k$ is spanned by every of its apartments. The restrictions of the Grassmann graph $\Gamma_k(V)$ to apartments of $[M, N]_k$ are isomorphic to the Johnson graph $J(l - m, k - m)$.

Theorem 3.8 ([Cooperstein, Kasikova and Shult (2005)]). *Let $1 < k < n-1$ and l, m be natural numbers satisfying $k < l \leq n$ and $0 \leq m < k$. Let also $\mathcal{X}$ be a subset of $\mathcal{G}_k(V)$ such that $\Gamma(\mathcal{X})$ is isomorphic to the Johnson graph $J(l - m, k - m)$ and every maximal clique of $\Gamma(\mathcal{X})$ is an independent subset of the Grassmann space $\mathfrak{G}_k(V)$. Then $\mathcal{X}$ is an apartment in a parabolic subspace of $\mathfrak{G}_k(V)$; in particular, this is an apartment of $\mathcal{G}_k(V)$ if $l = n$ and $m = 0$.*

The following example shows that the second condition in Theorem 3.8 (concerning the independence of maximal cliques) cannot be dropped.

Example 3.4. Let us take the following five vectors

$$(1,0,0,0,0)$$
$$(0,1,0,0,0)$$
$$(0,0,1,0,0)$$
$$(0,0,0,1,0)$$
$$(1,1,1,1,0)$$

in $\mathbb{R}^5$ and consider the subset of $\mathcal{G}_2(\mathbb{R}^5)$ consisting of all 2-dimensional linear subspaces spanned by pairs of these vectors. The restriction of the Grassmann graph $\Gamma_2(\mathbb{R}^5)$ to this subset is isomorphic to $J(5,2)$. However, it is not an apartment of $\mathcal{G}_2(\mathbb{R}^5)$.

Theorem 3.1 is a simple consequence of Theorem 3.8 and Proposition 3.5. Let f be a collineation of the Grassmann space $\mathfrak{G}_m(W)$ to a subspace of $\mathfrak{G}_k(V)$. We take any apartment $\mathcal{A} \subset \mathcal{G}_m(W)$. Then $f(\mathcal{A})$ satisfies the conditions of Theorem 3.8; hence it is an apartment in a parabolic subspace of $\mathfrak{G}_k(V)$. This parabolic subspace is spanned by $f(\mathcal{A})$ and we get the claim.

3.3.2 *Proof of Theorem 3.8*

If $l = k+1$ or $m = k-1$ then any two distinct elements of $\mathcal{X}$ are adjacent and the statement is trivial.

Suppose that $k+1 < l$ and $m < k-1$. Then $\mathcal{X}$ contains non-adjacent elements. Let $\mathcal{Y}$ be a maximal clique of $\Gamma(\mathcal{X})$. Then $\mathcal{Y}$ is contained in precisely one maximal clique of $\Gamma_k(V)$ (since $\mathcal{Y}$ is an independent subset of $\mathfrak{G}_k(V)$ containing more than 2 elements and the intersection of two distinct maximal cliques of $\Gamma_k(V)$ does not contain a triangle). We say that $\mathcal{Y}$ is a *star* or a *top* of $\mathcal{X}$ if the maximal clique of $\Gamma_k(V)$ containing $\mathcal{Y}$ is a star or a top, respectively.

We take an $(l-m)$-dimensional vector space W and any apartment $\mathcal{A} \subset \mathcal{G}_{k-m}(W)$. Let $f : \mathcal{A} \to \mathcal{X}$ be an isomorphism of $\Gamma(\mathcal{A})$ to $\Gamma(\mathcal{X})$. As in the proof of Theorem 3.2, one of the following possibilities is realized:

(A) stars go to stars and tops go to tops,
(B) stars go to tops and tops go to stars.

In the case (B), we consider the apartment $\mathcal{A}^* \subset \mathcal{G}_{l-k}(W^*)$ formed by the annihilators of the elements from $\mathcal{A}$; the mapping $U \to f(U^0)$ is an isomorphism of $\Gamma(\mathcal{A}^*)$ to $\Gamma(\mathcal{X})$ satisfying (A). By this reason, we restrict ourselves to the case (A).

Every top $\mathcal{Y} \subset \mathcal{X}$ consists of $k - m + 1$ elements; suppose that

$$\mathcal{Y} = \{S_1, \ldots, S_{k-m+1}\}.$$

Since this is an independent subset, the linear subspace

$$M_{\mathcal{Y}} := S_1 \cap \cdots \cap S_{k-m+1}$$

is m-dimensional, and for every $i \in \{1, \ldots, k - m + 1\}$

$$X_i := \bigcap_{j \neq i} S_j$$

is an $(m + 1)$-dimensional linear subspace. Denote by $\mathcal{B}_{\mathcal{Y}}$ the set formed by all X_i. This is a base of $[M_{\mathcal{Y}}, T]_{m+1}$, where T is the $(k + 1)$-dimensional linear subspace corresponding to $\mathcal{Y}$. Every element of $\mathcal{Y}$ is the sum of $k - m$ elements of $\mathcal{B}_{\mathcal{Y}}$.

Suppose that a top $\mathcal{Y}' \subset \mathcal{X}$ has a non-empty intersection with $\mathcal{Y}$. Then the $(k + 1)$-dimensional linear subspaces corresponding to $\mathcal{Y}$ and $\mathcal{Y}'$ are adjacent and $\mathcal{Y} \cap \mathcal{Y}'$ consists of unique element S. For every $U \in \mathcal{Y} \setminus \{S\}$ there exists unique $U' \in \mathcal{Y}' \setminus \{S\}$ adjacent with U (this follows from the fact that $\Gamma(\mathcal{X})$ is isomorphic to the Johnson graph $J(l - m, k - m)$). Since S, U, U' form a star-triangle, we have

$$S \cap U = S \cap U'.$$

This implies the following properties:

(1) $M_{\mathcal{Y}} = M_{\mathcal{Y}'}$,
(2) if $X \in \mathcal{B}_{\mathcal{Y}}$ is contained in S then it belongs to $\mathcal{B}_{\mathcal{Y}'}$; thus $\mathcal{B}_{\mathcal{Y}} \cap \mathcal{B}_{\mathcal{Y}'}$ consists of $k - m$ elements and there is a unique element of $\mathcal{B}_{\mathcal{Y}}$ which does not belong to $\mathcal{B}_{\mathcal{Y}'}$.

By connectedness of Grassmann spaces, $M_{\mathcal{Y}} = M_{\mathcal{Y}'}$ for any two tops $\mathcal{Y}, \mathcal{Y}' \subset \mathcal{X}$; in what follows this m-dimensional linear subspace will be denoted by M.

For every $S \in \mathcal{X}$ we denote by $\mathcal{B}(S)$ the union of all $\mathcal{B}_{\mathcal{Y}}$ such that $S \in \mathcal{Y}$. There are precisely $l - k$ distinct tops of $\mathcal{X}$ containing S and, by the property (2), the set $\mathcal{B}(S)$ consists of $l - m$ elements. Now, we show that $\mathcal{B}(S)$ coincides with $\mathcal{B}(U)$ if $S, U \in \mathcal{X}$ are adjacent.

There is unique top $\mathcal{Y} \subset \mathcal{X}$ containing S and U. By the definition, $\mathcal{B}_{\mathcal{Y}}$ is contained in both $\mathcal{B}(S)$ and $\mathcal{B}(U)$. Consider $X \in \mathcal{B}(S) \setminus \mathcal{B}_{\mathcal{Y}}$. Let $\mathcal{Y}' \subset \mathcal{X}$ be a top containing S and such that $X \in \mathcal{B}_{\mathcal{Y}'}$. We choose unique

$U' \in \mathcal{Y}'$ adjacent with U and denote by $\mathcal{Z}$ the top of $\mathcal{X}$ containing U and U'. Since X is not contained in S ($X \notin \mathcal{B}_\mathcal{Y}$), we have $X \subset U'$; by the property (2), this implies $X \in \mathcal{B}_\mathcal{Z} \subset \mathcal{B}(U)$. Therefore, $\mathcal{B}(S) \subset \mathcal{B}(U)$; the same arguments give the inverse inclusion.

By connectedness,

$$\mathcal{B}(S) = \mathcal{B}(U) \quad \forall \, S, U \in \mathcal{X};$$

denote this set by $\mathcal{B}$. Since $\mathcal{B}$ is formed by $l - m$ distinct $(m+1)$-dimensional linear subspaces containing M, the sum of all elements of $\mathcal{B}$ is a linear subspace N whose dimension is not greater than l; also we have $M \subset N$. On the other hand, every element of $\mathcal{X}$ is the sum of $k - m$ elements from $\mathcal{B}$; and every star of $\mathcal{X}$ is an independent subset consisting of $l - k + 1$ elements. This implies that $\dim N \geq l$ and N is l-dimensional. Then $\mathcal{B}$ is a base of $[M, N]_{m+1}$ and $\mathcal{X}$ is the associated apartment of $[M, N]_k$ (we get an apartment of $\mathcal{G}_k(V)$ if $l = n$ and $m = 0$).

3.4 Apartments preserving mappings

Let V and V' be n-dimensional vector spaces over division rings and $n \geq 3$. In this section we investigate apartments preserving mappings of $\mathcal{G}_k(V)$ to $\mathcal{G}_k(V')$. Recall that every apartments preserving mapping is injective.

3.4.1 *Results*

All collineations of $\mathfrak{G}_k(V)$ to $\mathfrak{G}_k(V')$ (if they exist) are apartments preserving (since they are induced by semilinear isomorphisms). Conversely, we have the following.

Theorem 3.9 (M. Pankov). *Every apartments preserving bijection of $\mathcal{G}_k(V)$ to $\mathcal{G}_k(V')$ is a collineation of $\mathfrak{G}_k(V)$ to $\mathfrak{G}_k(V')$.*

Remark 3.9. If V is a vector space over a field then every base $\{x_1, \ldots, x_n\}$ of V defines the *regular* base of $\Pi_{\wedge^k V}$ consisting of all

$$\langle x_{i_1} \wedge \cdots \wedge x_{i_k} \rangle.$$

Clearly, the projective space $\Pi_{\wedge^k V}$ has non-regular bases. The Grassmann embedding g_k (Remark 3.1) establishes a one-to-one correspondence between apartments of $\mathcal{G}_k(V)$ and regular bases of $\Pi_{\wedge^k V}$. Let f be a bijective transformation of the projective space $\Pi_{\wedge^k V}$ preserving the family of regular bases (f and f^{-1} map regular bases to regular bases). A point of

$\Pi_{\wedge^k V}$ is contained in a regular base if and only if it belongs to the image of $\mathcal{G}_k(V)$. Thus f transfers the image of $\mathcal{G}_k(V)$ to itself and the restriction of f to this subset can be identified with an apartments preserving bijective transformation of $\mathcal{G}_k(V)$.

Let $l : V \to V'$ be a semilinear embedding. For every $k \in \{1, \ldots, n-1\}$ the mapping

$$(l)_k : \mathcal{G}_k(V) \to \mathcal{G}_k(V')$$

(Subsection 1.3.2) is an apartments preserving embedding of $\mathfrak{G}_k(V)$ in $\mathfrak{G}_k(V')$. Recall that $(l)_k$ is bijective if and only if l is a semilinear isomorphism (Proposition 1.11).

Let $u : V \to V'^*$ be a semilinear embedding. By duality, we can identify $(u)_k$ with the embedding of $\mathfrak{G}_k(V)$ in $\mathfrak{G}_{n-k}(V')$ sending each S to $\langle u(S) \rangle^0$. In the case when $n = 2k$, this is an apartments preserving embedding of $\mathfrak{G}_k(V)$ in $\mathfrak{G}_k(V')$.

Theorem 3.10 (M. Pankov). *Let $1 < k < n-1$. Then every apartments preserving mapping of $\mathcal{G}_k(V)$ to $\mathcal{G}_k(V')$ is the embedding of $\mathfrak{G}_k(V)$ in $\mathfrak{G}_k(V')$ induced by a semilinear embedding of V in V' or a semilinear embedding of V in V'^*; the second possibility can be realized only in the case when $n = 2k$.*

If $k = 1, n-1$ then there exist apartments preserving mappings of $\mathcal{G}_k(V)$ to $\mathcal{G}_k(V')$ which cannot be induced by semilinear mappings.

Example 3.5 ([Huang and Kreuzer (1995)]). Let us consider the injective mapping $\alpha : \mathbb{R} \to \mathcal{G}_1(\mathbb{R}^n)$ defined by the formula

$$t \to \langle (t, t^2, \ldots, t^n) \rangle.$$

For any distinct $t_1, \ldots, t_n \in \mathbb{R} \setminus \{0\}$ we have

$$\begin{vmatrix} t_1 & \cdots & t_1^n \\ \vdots & \ddots & \vdots \\ t_n & \cdots & t_n^n \end{vmatrix} \neq 0.$$

Hence any subset of $\alpha(\mathbb{R})$ consisting of n distinct points is a base of the projective space $\Pi_{\mathbb{R}^n}$; in particular, any three points of $\alpha(\mathbb{R})$ are non-collinear. The sets $\mathbb{R}$ and $\mathcal{G}_1(\mathbb{R}^n)$ have the same cardinality and we consider any bijection of $\mathcal{G}_1(\mathbb{R}^n)$ to $\alpha(\mathbb{R})$. This mapping transfers bases of $\Pi_{\mathbb{R}^n}$ to bases of $\Pi_{\mathbb{R}^n}$, but it is not induced by a semilinear mapping (since the images of three distinct collinear points are non-collinear).

In the case when $1 < k < n - 1$, Theorem 3.9 is a direct consequence of Theorem 3.10 and Proposition 1.11.

Let f be an apartments preserving bijection of $\mathcal{G}_k(V)$ to $\mathcal{G}_k(V')$. Suppose that $k = 1$. Then f maps bases of Π_V to bases of $\Pi_{V'}$. Since three points of a projective space are non-collinear if and only if they are contained in a certain base of this space, the mapping f sends each triple of non-collinear points to non-collinear points. This implies that the inverse mapping transfers triples of collinear points to collinear points. Hence f^{-1} is a semicollineation of $\Pi_{V'}$ to Π_V and Corollary 1.3 gives the claim. The case $k = n - 1$ is similar, by Exercise 1.11.

Remark 3.10. Theorem 3.9 was first proved in [Pankov (2002)]. Our proof of Theorem 3.10 is a modification of the proof given in [Pankov 2 (2004); Pankov 3 (2004)].

3.4.2 *Proof of Theorem 3.10: First step*

Let $B = \{x_1, \ldots, x_n\}$ be a base of V and $\mathcal{A}$ be the associated apartment of $\mathcal{G}_k(V)$. Throughout the subsection we suppose that $1 < k \le n - k$.

We write $\mathcal{A}(+i)$ and $\mathcal{A}(-i)$ for the sets consisting of all elements of $\mathcal{A}$ which contain x_i and do not contain x_i, respectively. If S is a linear subspace spanned by a subset of B then we denote by $\mathcal{A}(S)$ the set of all elements of $\mathcal{A}$ incident with S; in the case when S is spanned by x_i and x_j, we will write $\mathcal{A}(+i, +j)$ instead of $\mathcal{A}(S)$. Clearly, $\mathcal{A}(S)$ coincides with a certain $\mathcal{A}(-i)$ if S is $(n - 1)$-dimensional.

A subset of $\mathcal{A}$ is called *exact* if it is contained only in one apartment of $\mathcal{G}_k(V)$; otherwise, it is said to be *inexact*. It is trivial that $\mathcal{R} \subset \mathcal{A}$ is exact if the intersection of all $S \in \mathcal{R}$ containing x_i coincides with $\langle x_i \rangle$ for every i.

Lemma 3.2. *A subset* $\mathcal{R} \subset \mathcal{A}$ *is exact if and only if for each i the intersection of all $S \in \mathcal{R}$ containing x_i coincides with $\langle x_i \rangle$.*

Proof. Suppose that for a certain number i there exists $j \ne i$ such that x_j belongs to every $S \in \mathcal{R}$ containing x_i. We choose a vector $x = ax_i + bx_j$ with $a, b \ne 0$. Then

$$(B \setminus \{x_i\}) \cup \{x\}$$

is a base of V which defines another apartment of $\mathcal{G}_k(V)$ containing $\mathcal{R}$, thus $\mathcal{R}$ is inexact. If there are no elements of $\mathcal{R}$ containing x_i then we can take any vector x which is not a scalar multiple of x_i and is not contained in the linear subspace spanned by $B \setminus \{x_i\}$. $\qquad\square$

By Lemma 3.2,
$$\mathcal{A}(-i) \cup \mathcal{A}(+i, +j), \ i \neq j,$$
is an inexact subset.

Lemma 3.3. *If $\mathcal{R}$ is a maximal inexact subset of $\mathcal{A}$ then*
$$\mathcal{R} = \mathcal{A}(-i) \cup \mathcal{A}(+i, +j)$$
for some distinct i, j.

Proof.   For each i there is an element of $\mathcal{R}$ containing x_i (if all elements of $\mathcal{R}$ do not contain x_i then $\mathcal{R}$ is a subset of the non-maximal inexact subset $\mathcal{A}(-i)$ which contradicts the fact that our inexact subset is maximal). By Lemma 3.2, there exist distinct i and j such that x_j belongs to all elements of $\mathcal{R}$ containing x_i. Then every $S \in \mathcal{R}$ is an element of $\mathcal{A}(-i)$ or $\mathcal{A}(+i, +j)$ and
$$\mathcal{R} \subset \mathcal{A}(-i) \cup \mathcal{A}(+i, +j).$$
Since $\mathcal{R}$ is a maximal inexact subset, we have the inverse inclusion.   □

A subset $\mathcal{R} \subset \mathcal{A}$ is said to be *complement* if $\mathcal{A} \setminus \mathcal{R}$ is a maximal inexact subset. In this case, Lemma 3.3 implies the existence of distinct i, j such that
$$\mathcal{A} \setminus \mathcal{R} = \mathcal{A}(-i) \cup \mathcal{A}(+i, +j).$$
Then
$$\mathcal{R} = \mathcal{A}(+i) \cap \mathcal{A}(-j);$$
in what follows this complement subset will be denoted by $\mathcal{A}(+i, -j)$.

We say that distinct complement subsets
$$\mathcal{R}_1 = \mathcal{A}(+i_1, -j_1), \dots, \mathcal{R}_k = \mathcal{A}(+i_k, -j_k)$$
form a *regular collection* if their intersection is a one-element set. In the general case,
$$\mathcal{R}_1 \cap \cdots \cap \mathcal{R}_k = \mathcal{A}(M) \cap \mathcal{A}(N),$$
where M and N are linear subspaces spanned by subsets of B and
$$\dim M = |\{i_1, \dots, i_k\}| \leq k \leq n - k \leq n - |\{j_1, \dots, j_k\}| = \dim N$$
(note that some of $i_1, \dots, i_k$ or $j_1, \dots, j_k$ can be coincident). This intersection is not empty if and only if
$$\{i_1, \dots, i_k\} \cap \{j_1, \dots, j_k\} = \emptyset; \tag{3.4}$$
in this case, M is contained in N.

Example 3.6. If (3.4) holds and $i_1, \dots, i_k$ are distinct then
$$\mathcal{R}_1 \cap \cdots \cap \mathcal{R}_k = \{\langle x_{i_1}, \dots, x_{i_k} \rangle\}$$
and the collection is regular.

Example 3.7. Suppose that $n = 2k$, the condition (3.4) holds, and $j_1, \ldots, j_k$ are distinct. Then

$$\mathcal{R}_1 \cap \cdots \cap \mathcal{R}_k = \{\langle x_{i'_1}, \ldots, x_{i'_k} \rangle\},$$

where

$$\{i'_1, \ldots, i'_k\} = \{1, \ldots, n\} \setminus \{j_1, \ldots, j_k\}.$$

The collection is regular.

Lemma 3.4. *The collection* $\mathcal{R}_1, \ldots, \mathcal{R}_k$ *is regular if and only if* (3.4) *holds and one of the following possibilities is realized:*

(A) $i_1, \ldots, i_k$ *are distinct,*
(B) $n = 2k$ *and* $j_1, \ldots, j_k$ *are distinct.*

Proof. If (3.4) is fulfilled and the conditions (A) and (B) both do not hold then

$$\dim M < k < \dim N$$

and $\mathcal{A}(M) \cap \mathcal{A}(N)$ contains more than one element. $\square$

A collection of $k - 1$ distinct complement subsets $\mathcal{R}_1, \ldots, \mathcal{R}_{k-1} \subset \mathcal{A}$ is said to be *regular* if it can be extended to a regular collection of k distinct complement subsets; in other words, there exists a complement subset $\mathcal{R}_k \subset \mathcal{A}$ such that

$$\mathcal{R}_k \neq \mathcal{R}_1, \ldots, \mathcal{R}_{k-1} \text{ and } \mathcal{R}_1, \ldots, \mathcal{R}_{k-1}, \mathcal{R}_k$$

is a regular collection.

The adjacency relation can be characterized in terms of regular collections of complement subsets.

Lemma 3.5. *The following conditions are equivalent:*

(1) $S, U \in \mathcal{A}$ *are adjacent,*
(2) *there exists a regular collection of* $k - 1$ *distinct complement subsets of* $\mathcal{A}$ *such that each element of this collection contains S and U.*

Proof. (1) $\Longrightarrow$ (2). Suppose that S and U are adjacent and

$$S \cap U = \langle x_{i_1}, \ldots, x_{i_{k-1}} \rangle.$$

We choose

$$x_i \in S \setminus U \text{ and } x_j \notin S + U.$$

Then

$$\mathcal{A}(+i_1, -j) \cap \cdots \cap \mathcal{A}(+i_{k-1}, -j) \cap \mathcal{A}(+i, -j) = \{S\}.$$

Hence these complement subsets form a regular collection and the complement subsets

$$\mathcal{A}(+i_1, -j), \ldots, \mathcal{A}(+i_{k-1}, -j)$$

are as required.

(2) $\Longrightarrow$ (1). Conversely, suppose that each element of a regular collection

$$\mathcal{A}(+i_1, -j_1), \ldots, \mathcal{A}(+i_{k-1}, -j_{k-1})$$

contains S and U. By definition, this collection can be extended to a regular collection of k distinct complement subsets; and it follows from Lemma 3.4 that one of the following possibilities is realized:

(A) $i_1, \ldots, i_{k-1}$ are distinct,
(B) $n = 2k$ and $j_1, \ldots, j_{k-1}$ are distinct.

Since $x_{i_1}, \ldots, x_{i_{k-1}}$ belong to $S \cap U$, (A) guarantees that the dimension of $S \cap U$ is equal to $k - 1$. The vectors $x_{j_1}, \ldots, x_{j_{k-1}}$ do not belong to $S + U$ and (B) implies that $S + U$ is $(k + 1)$-dimensional. In each of these cases, S and U are adjacent. $\qquad\square$

Proposition 3.6. *Let $1 < k \le n - k$ and $f : \mathcal{G}_k(V) \to \mathcal{G}_k(V')$ be an apartments preserving mapping. Then f is adjacency preserving: two elements of $\mathcal{G}_k(V)$ are adjacent if and only if their images are adjacent.*

Proof.   First of all we recall that f is injective. Let S and U be k-dimensional linear subspaces of V and $\mathcal{A}$ be an apartment of $\mathcal{G}_k(V)$ containing them. Then $f(\mathcal{A})$ is an apartment of $\mathcal{G}_k(V')$ and f maps inexact subsets of $\mathcal{A}$ to inexact subsets of $f(\mathcal{A})$; indeed, if a subset $\mathcal{R}$ is contained in two distinct apartments of $\mathcal{G}_k(V)$ then $f(\mathcal{R})$ is contained in their images which are distinct apartments of $\mathcal{G}_k(V')$. Since $\mathcal{A}$ and $f(\mathcal{A})$ have the same number of inexact subsets, every inexact subset of $f(\mathcal{A})$ is the image of a certain inexact subset of $\mathcal{A}$. This implies that an inexact subset of $\mathcal{A}$ is maximal if and only if its image is a maximal inexact subset of $f(\mathcal{A})$. Therefore, a subset $\mathcal{R} \subset \mathcal{A}$ is complement if and only if $f(\mathcal{R})$ is a complement subset of $f(\mathcal{A})$; moreover, $f|_{\mathcal{A}}$ and the inverse mapping transfer regular collections of complement subsets to regular collections (this follows immediately from the definition of a regular collection). By Lemma 3.5, S and U are adjacent if and only if $f(S)$ and $f(U)$ are adjacent. $\qquad\square$

3.4.3 *Proof of Theorem 3.10: Second step*

By duality, every apartments preserving mapping $f : \mathcal{G}_k(V) \to \mathcal{G}_k(V')$ can be identified with an apartments preserving mapping of $\mathcal{G}_{n-k}(V^*)$ to $\mathcal{G}_{n-k}(V'^*)$. Suppose that the latter mapping is induced by a semilinear embedding $u : V^* \to V'^*$ and consider the mapping

$$g : \mathcal{G}(V) \to \mathcal{G}(V')$$

$$S \to \langle u(S^0) \rangle^0.$$

The restriction of g to $\mathcal{G}_k(V)$ coincides with f. The mapping g is dimension preserving and we have

$$S \subset U \implies g(S) \subset g(U)$$

for all $S, U \in \mathcal{G}(V)$. By Theorem 1.4, the restriction of g to $\mathcal{G}_1(V)$ is induced by a semilinear injection $l : V \to V'$. Since this restriction is apartments preserving, l is a semilinear embedding. For every linear subspace $S \subset V$ we choose $P_1, \ldots, P_m \in \mathcal{G}_1(S)$, $m = \dim S$ such that

$$S = P_1 + \cdots + P_m.$$

Then

$$\langle l(S) \rangle = \langle l(P_1) \rangle + \cdots + \langle l(P_m) \rangle = g(P_1) + \cdots + g(P_m) \subset g(S)$$

which implies that $\langle l(S) \rangle$ coincides $g(S)$ (since the dimension of these linear subspaces is equal to m). Thus g is induced by l and $f = (l)_k$.

Therefore, we can prove the theorem only in the case when $k \leq n - k$. Let $1 < k \leq n - k$ and f be an apartments preserving mapping of $\mathcal{G}_k(V)$ to $\mathcal{G}_k(V')$.

By Proposition 3.6, f is an adjacency preserving injection. Hence it transfers stars and tops to subsets of stars or tops. For every star there is an apartment intersecting this star in a set consisting of $n - k + 1$ elements; on the other hand, the intersections of tops with apartments contain at most $k + 1$ elements. Since $k \leq n - k$, the image of a star can be a subset of a top only in the case when $n = 2k$. In particular, stars go to subsets of stars if $k < n - k$.

By the same reason, the image of every maximal singular subspace of $\mathfrak{G}_k(V)$ is contained in precisely one maximal singular subspace of $\mathfrak{G}_k(V')$. Indeed, the intersection of two distinct maximal singular subspaces is empty, or a single point, or a line; thus it intersects apartments in subsets containing at most two elements.

Distinct maximal singular subspaces of $\mathfrak{G}_k(V)$ go to subsets of distinct maximal singular subspaces of $\mathfrak{G}_k(V')$ (otherwise there exist non-adjacent elements of $\mathcal{G}_k(V)$ whose images are adjacent).

Suppose that f transfers stars to subsets of stars. Then, as in the proof of Theorem 3.2, f induces an injective mapping

$$f_{k-1} : \mathcal{G}_{k-1}(V) \to \mathcal{G}_{k-1}(V')$$

such that

$$f([S\rangle_k) \subset [f_{k-1}(S)\rangle_k$$

for all $S \in \mathcal{G}_{k-1}(V)$. Now we show that f_{k-1} is apartments preserving.

Consider a base $B \subset V$ and the associated apartment $\mathcal{A}_k \subset \mathcal{G}_k(V)$. Let B' be one of the bases of V' associated with the apartment $\mathcal{A}'_k := f(\mathcal{A}_k)$. Let also

$$\mathcal{A}_{k-1} \subset \mathcal{G}_{k-1}(V) \quad \text{and} \quad \mathcal{A}'_{k-1} \subset \mathcal{G}_{k-1}(V')$$

be the apartments defined by the bases B and B', respectively. Every $S \in \mathcal{A}_{k-1}$ is the intersection of two adjacent $U_1, U_2 \in \mathcal{A}_k$ and

$$f_{k-1}(S) = f(U_1) \cap f(U_2) \in \mathcal{A}'_{k-1}.$$

Since the mapping f_{k-1} is injective and $|\mathcal{A}_{k-1}| = |\mathcal{A}'_{k-1}| < \infty$, we get $f_{k-1}(\mathcal{A}_{k-1}) = \mathcal{A}'_{k-1}$.

Step by step, we construct a sequence of apartments preserving mappings

$$f_i : \mathcal{G}_i(V) \to \mathcal{G}_i(V'), \quad i = k, \ldots, 1,$$

such that $f_k = f$,

$$f_i([S\rangle_i) \subset [f_{i-1}(S)\rangle_i$$

for all $S \in \mathcal{G}_{i-1}(V)$ and

$$f_{i-1}(\langle U]_{i-1}) \subset \langle f_i(U)]_{i-1}$$

for all $U \in \mathcal{G}_i(V)$ if $i > 1$. By Theorem 1.4, f_1 is induced by a semilinear injection $l : V \to V'$. Since f_1 is apartments preserving, l is a semilinear embedding.

We prove that $f_i = (l)_i$ induction by i. If f_{i-1} is induced by l then

$$f_{i-1}(\langle U]_{i-1}) \subset \langle f_i(U)]_{i-1} \cap \langle \langle l(U) \rangle]_{i-1}$$

for every $U \in \mathcal{G}_i(V)$. The latter intersection contains more that one element which implies that $f_i(U)$ coincides with $\langle l(U) \rangle$.

In the case when the image of a certain star of $\mathcal{G}_k(V)$ is contained in a top, we show that each star goes to a subset of a top (the standard arguments from the proof of Theorem 3.2). Since $n = 2k$, we can identify f with an apartments preserving mapping of $\mathcal{G}_k(V)$ to $\mathcal{G}_k(V'^*)$ which sends stars to subsets of stars. The latter mapping is induced by a semilinear embedding of V in V'^*.

3.5 Grassmannians of exchange spaces

3.5.1 *Exchange spaces*

We say that a linear space $\Pi = (P, \mathcal{L})$ satisfies the *exchange axiom*, or simple, Π is an *exchange* space if for every subset $X \subset P$ and any points $p, q \in P \setminus \langle X \rangle$

$$p \in \langle X, q \rangle \implies q \in \langle X, p \rangle.$$

This axiom holds for projective and affine spaces.

Theorem 3.11. *In an exchange space every independent subset can be extended to a base of this space, and any two bases of an exchange space have the same cardinality.*

Proof. See §8 in [Karzel, Sörensen and Windelberg (1973)]. □

Every semicollineation between projective spaces of the same finite dimension is a collineation (Corollary 1.3). The same holds for exchange spaces.

Theorem 3.12 ([Kreuzer (1996)]). *Let $\Pi = (P, \mathcal{L})$ and $\Pi' = (P', \mathcal{L}')$ be finite-dimensional exchange spaces such that*

$$\dim \Pi \leq \dim \Pi'. \tag{3.5}$$

Then every semicollineation of Π to Π' is a collineation.

Proof. Let f be a semicollineation of Π to Π'. Consider a subset $X \subset P$ containing at least two points. First, we prove induction by i that

$$f([X]_i) \subset [f(X)]_i$$

(Section 2.5). For $i = 1$ this is trivial; if the inclusion holds for $i = k - 1$ then

$$f([X]_k) = f([[X]_{k-1}]_1) \subset [f([X]_{k-1})]_1 \subset [[f(X)]_{k-1}]_1 = [f(X)]_k.$$

By Proposition 2.6, we have

$$f(\langle X \rangle) = \bigcup_{i=1}^{\infty} f([X]_i) \subset \bigcup_{i=1}^{\infty} [f(X)]_i = \langle f(X) \rangle.$$

Therefore, if B is a base of Π then

$$f(\langle B \rangle) = f(P) = P' \subset \langle f(B) \rangle$$

and Π' is spanned by $f(B)$. The inequality (3.5) guarantees that $f(B)$ is a base of Π'. Thus our spaces have the same dimension and f transfers bases of Π to bases of Π'. This implies that f maps any triple of non-collinear points to non-collinear points (since three distinct points are non-collinear if and only if there is a base containing them, Theorem 3.11). Then f^{-1} sends triples of collinear points to collinear points, hence f^{-1} is a semicollineation of Π' to Π.   □

Now suppose that our exchange spaces have the same finite dimension and consider a mapping $f : P \to P'$ which sends bases of Π to bases of Π'. Then f is injective and transfers any triple of non-collinear points to non-collinear points (standard arguments). In the case when f is bijective, the inverse mapping is a semicollineation of Π' to Π. Theorem 3.12 gives the following.

Corollary 3.3. *Let $\Pi = (P, \mathcal{L})$ and $\Pi' = (P', \mathcal{L}')$ be exchange spaces of the same finite dimension. Then every bijection of P to P' sending bases of Π to bases of Π' is a collineation of Π to Π'.*

The following example shows that the condition (3.5) in Theorem 3.12 cannot be dropped.

Example 3.8 ([Kreuzer (1996)]). Let S be a plane in a 3-dimensional projective space $\Pi = (P, \mathcal{L})$. Consider the following set of lines

$$\mathcal{L}' := \{S\} \cup \{\, L \in \mathcal{L} \ : \ L \not\subset S \,\};$$

in other words, we remove all lines contained in S and add S as a line. An easy verification shows that $\Pi' := (P, \mathcal{L}')$ is an exchange plane. The identity transformation of P is a semicollineation of Π to Π' which is not a collineation.

Remark 3.11. A more complicated example of a semicollineation of a 4-dimensional projective space to a non-Desarguesian projective plane was given in [Ceccherini (1967)].

3.5.2 *Grassmannians*

Let $\Pi = (P, \mathcal{L})$ be an exchange space of finite dimension n. For every number $k \in \{0, \ldots, n-1\}$ we denote by $\mathcal{G}_k(\Pi)$ the Grassmannian consisting of all k-dimensional subspaces of Π. It follows from Theorem 3.11 that any two incident k-dimensional subspaces of Π are coincident; note that for an arbitrary linear space the same holds only in the case when $k = 0, 1$.

Two elements of the Grassmannian $\mathcal{G}_k(\Pi)$ are said to be *adjacent* if their intersection is $(k - 1)$-dimensional.

Lemma 3.6. *If* $S, U \in \mathcal{G}_k(\Pi)$ *are adjacent then the subspace* $\langle S, U \rangle$ *is* $(k + 1)$-*dimensional.*

Proof. We take any base X of $S \cap U$ and any points $p \in S \setminus U$ and $q \in U \setminus S$. The exchange axiom guarantees that $X \cup \{p\}$, $X \cup \{q\}$, and $X \cup \{p, q\}$ are bases of S, U, and $\langle S, U \rangle$ (respectively). $\quad\square$

Remark 3.12. It was noted in Subsection 3.2.2 that there exist pairs of non-adjacent elements of $\mathcal{G}_k(\Pi)$ which span $(k + 1)$-dimensional subspaces.

The *Grassmann graph* $\Gamma_k(\Pi)$ is the graph whose vertex set is $\mathcal{G}_k(\Pi)$ and whose edges are pairs of adjacent subspaces. It is not difficult to prove that this graph is connected and every star is a maximal clique (stars and tops are defined as for Grassmannians of finite-dimensional vector spaces).

Proposition 3.7. *Every maximal clique of* $\Gamma_k(\Pi)$ *is a star or a subset of a top.*

Proof. Similar to the proof of Proposition 3.2, see [Pankov (2006)]. $\quad\square$

For index one Grassmann graphs we have an analogue of Chow's theorem (Theorem 3.4); note that the exchange axiom is not required. For larger indices such kind results are unknown.

Consider a certain base B of Π. The set consisting of all k-dimensional subspaces spanned by subsets of B is called the *base subset* of $\mathcal{G}_k(\Pi)$ associated with (defined by) the base B. In the case when Π is the projective space of an $(n + 1)$-dimensional vector space V, the Grassmannian $\mathcal{G}_k(\Pi)$ can be identified with $\mathcal{G}_{k+1}(V)$ and base subsets are apartments.

Lemma 3.7. *For any adjacent* $S, U \in \mathcal{G}_k(\Pi)$ *there is a base subset of* $\mathcal{G}_k(\Pi)$ *containing* S *and* U.

Proof. There exists an independent subset $Y \subset P$ such that S and U are spanned by subsets of Y (see the proof of Lemma 3.6). We extend Y to a base of Π; the associated base subset of $\mathcal{G}_k(\Pi)$ contains S and U. $\quad\square$

Remark 3.13. Two k-dimensional subspaces S and U are contained in a base subset of $\mathcal{G}_k(\Pi)$ if and only if

$$\dim\langle S, U \rangle = 2k - \dim(S \cap U),$$

see [Pankov (2006)]. In the general (non-projective) case, this formula does not hold and there exist pairs of k-dimensional subspaces which are not contained in base subsets of $\mathcal{G}_k(\Pi)$.

Theorem 3.13 ([Pankov (2006)]). *Let $\Pi = (P, \mathcal{L})$ and $\Pi' = (P', \mathcal{L}')$ be exchange spaces of the same finite dimension $n \geq 2$ such that each line of Π and Π' contains at least 3 points. Suppose that $f : \mathcal{G}_k(\Pi) \to \mathcal{G}_k(\Pi')$ is an injection sending base subsets to base subsets. Then f maps adjacent elements to adjacent elements.*

Proof. The case $k = 0$ is trivial (any two distinct points are adjacent). If $k = n - 1$ then the required statement is a direct consequence of the following observation: two distinct $(n-1)$-dimensional subspaces of Π or Π' are adjacent if and only if there is a base subset containing them.

Let $0 < k < n - 1$. Consider a base $B = \{p_1, \ldots, p_{n+1}\}$ of Π and the associated base subset $\mathcal{B} \subset \mathcal{G}_k(\Pi)$. We write $\mathcal{B}(+i)$ and $\mathcal{B}(-i)$ for the sets consisting of all elements of $\mathcal{B}$ which contain p_i and do not contain p_i, respectively. We define exact, inexact, and complement subsets of $\mathcal{B}$ as for apartments in Grassmannians of finite-dimensional vector spaces.

Using the fact that every line contains at least three points, we establish the direct analogue of Lemma 3.2: a subset $\mathcal{R} \subset \mathcal{B}$ is exact if and only if for every i the intersection of all $S \in \mathcal{R}$ containing p_i coincides with p_i. We describe maximal inexact subsets of $\mathcal{B}$ (cf. Lemma 3.3) and show that every complement subset of $\mathcal{B}$ is

$$\mathcal{B}(+i, -j) := \mathcal{B}(+i) \cap \mathcal{B}(-j), \quad i \neq j.$$

Now define

$$m := \min\{k, n - k - 1\}.$$

We say that $m + 1$ distinct complement subsets

$$\mathcal{B}(+i_1, -j_1), \ldots, \mathcal{B}(+i_{m+1}, -j_{m+1})$$

form a *regular collection* if their intersection is a one-element set. In this case, one of the following three possibilities is realized (an analogue of Lemma 3.4):

(1) $m = k$ and $i_1, \ldots, i_{k+1}$ are distinct,
(2) $m = k = n - k - 1$ and $i_1, \ldots, i_{k+1}$ or $j_1, \ldots, j_{k+1}$ are distinct,
(3) $m = n - k - 1$ and $j_1, \ldots, j_{n-k}$ are distinct.

A collection of m distinct complement subsets is said to be *regular* if it can be extended to a regular collection of $m + 1$ distinct complement subsets. We have the standard characterization of the adjacency relation in terms of complement subsets (cf. Lemma 3.5); in other words, the following conditions are equivalent:

- $S, U \in \mathcal{B}$ are adjacent,
- there exists a regular collection of m distinct complement subsets of $\mathcal{B}$ such that each element of this collection contains S and U.

Now consider adjacent k-dimensional subspaces S and U of Π. By Lemma 3.7, there is a base subset $\mathcal{B} \subset \mathcal{G}_k(\Pi)$ containing them. The injectivity of f guarantees that inexact subsets of $\mathcal{B}$ go to inexact subsets of $f(\mathcal{B})$. As in the proof of Theorem 3.10, we establish that a subset of $\mathcal{B}$ is complement if and only if its image is a complement subset of $f(\mathcal{B})$. Therefore, two elements of $\mathcal{B}$ are adjacent if and only if the same holds for their images; in particular, $f(S)$ and $f(U)$ are adjacent. $\qquad\square$

Remark 3.14. In the general case, there are pairs of k-dimensional subspaces which are not contained in base subsets of $\mathcal{G}_k(\Pi)$; thus we cannot assert that a mapping of $\mathcal{G}_k(\Pi)$ to $\mathcal{G}_k(\Pi')$ sending base subsets to base subsets is injective. By the same reason, the pre-images of adjacent elements of $\mathcal{G}_k(\Pi')$ need not to be adjacent, see Example 3.9.

Theorem 3.14 ([Pankov (2006)]). *Let $\Pi = (P, \mathcal{L})$ and $\Pi' = (P', \mathcal{L}')$ be as in the previous theorem and f be a bijection of $\mathcal{G}_k(\Pi)$ to $\mathcal{G}_k(\Pi')$ such that f and f^{-1} map base subsets to base subsets. Then f is an isomorphism of $\Gamma_k(\Pi)$ to $\Gamma_k(\Pi')$; in the case when $2k + 1 < n$, it is induced by a collineation of Π to Π'.*

Proof. By Theorem 3.13, f is an isomorphism of $\Gamma_k(\Pi)$ to $\Gamma_k(\Pi')$. Therefore, f and f^{-1} map maximal cliques to maximal cliques. Every maximal clique is a star or a subset of a top (Proposition 3.7). In the case when $2k + 1 < n$, stars go to stars (for every star there is a base subset intersecting this star in a subset consisting of $n - k + 1$ elements, the intersections of base subsets with tops contain at most $k + 2$ elements). Thus f induces a bijection

$$f_{k-1} : \mathcal{G}_{k-1}(\Pi) \to \mathcal{G}_{k-1}(\Pi').$$

As in the proof of Theorem 3.10, we show that f_{k-1} maps base subsets to base subsets. The inverse mapping is induced by f^{-1}, hence it also sends base subsets to base subsets. We construct a sequence of bijections

$$f_i : \mathcal{G}_i(\Pi) \rightarrow \mathcal{G}_i(\Pi'), \quad i = k, \ldots, 0,$$

where $f_k = f$ and each f_i together with the inverse mapping send base subsets to base subsets. Then f_0 is a collineation of Π to Π' (Corollary 3.3) and an easy verification shows that f is induced by f_0. □

Remark 3.15. Since dual principle does not work for exchange spaces, we cannot apply our method to the case when $2k + 1 \geq n$. However, if $n = 3$, $k = 1$ and f is as in Theorem 3.14 then, by Theorem 3.4, one of the following possibilities is realized:

- f is induced by a collineation of Π to Π';
- Π and Π' both are projective spaces and f is induced by a collineation of Π to the projective space dual to Π'.

Example 3.9. Suppose that Π' is a projective space and Π is the linear space obtained from Π' by removing a certain point p. It is not difficult to prove that Π is an exchange space and $\dim \Pi = \dim \Pi'$. Consider the bijection of $\mathcal{L}$ to $\mathcal{L}'$ which sends every $L \in \mathcal{L}$ to the line of Π' containing L (this mapping is induced by the natural embedding of Π in Π'). This bijection transfers base subsets to base subsets. The inverse mapping does not satisfy this condition, since it is not adjacency preserving (lines of Π' passing through the point p go to non-intersecting lines of Π).

Remark 3.16. Every strong embedding of Π in Π' (an embedding sending independent subsets to independent subsets) induces an injection of $\mathcal{G}_k(\Pi)$ to $\mathcal{G}_k(\Pi')$ which transfers base subsets to base subsets. Using the arguments given above, we can show that every injection $f : \mathcal{L} \rightarrow \mathcal{L}'$ transferring base subsets to base subsets is induced by a strong embedding of Π in Π' [Pankov (2006)]. In contrast to the projective case, the assumption that f is bijective does not guarantee that this embedding is a collineation (Example 3.9).

3.6 Matrix geometry and spine spaces

Let V be an n-dimensional vector space over a division ring. Let also W be an $(n - k)$-dimensional linear subspace of V. Consider the set $\mathcal{X} \subset \mathcal{G}_k(V)$ consisting of all complements of W. There is a one-to-one correspondence between elements of $\mathcal{X}$ and all $k \times (n - k)$-matrices over the associated division ring: we take any $M \in \mathcal{X}$ and fix bases in M and W, then every

$S \in \mathcal{X}$ can be considered as the graph of a linear mapping $l : M \to W$, and we identify S with the matrix of this linear mapping in the fixed bases. The distance between two matrices A and B is defined as the rank of $A - B$; it is equal to the distance between the corresponding elements of $\mathcal{X}$. Information concerning adjacency preserving transformations of the set of rectangular matrices can be found in [Wan (1996)] (Theorem 3.4).

The construction considered above has a natural generalization [Prażmowski (2001)]. We fix a linear subspace $W \subset V$ and denote by $\mathcal{F}_{k,m}(W)$ the set consisting of all $S \in \mathcal{G}_k(V)$ satisfying

$$\dim(W \cap S) = m$$

(we assume that m is not greater than $\dim W$ and k which guarantees that the set $\mathcal{F}_{k,m}(W)$ is not empty). The restriction of the Grassmann space $\mathfrak{G}_k(V)$ to this set (Subsection 1.2.1) is a partial linear space. Partial linear spaces of such kind are called *spine spaces*. In the case when $m = 0$ and $\dim W = n - k$, we get the geometry of rectangular matrices. If m is equal to $\dim W$ or k then our spine space is a parabolic subspace of $\mathfrak{G}_k(V)$.

Suppose that $1 < k < n - 1$ and the line

$$[M, N]_k, \quad M \in \mathcal{G}_{k-1}(V), \; N \in \mathcal{G}_{k+1}(V),$$

induces a line in the spine space (the intersection of $[M, N]_k$ with $\mathcal{F}_{k,m}(W)$ contains at least two points). In the general case, there are the following three possibilities:

(τ) $\dim(M \cap W) = m$, $\dim(N \cap W) = m + 1$;
(α) $\dim(M \cap W) = \dim(N \cap W) = m$;
(ω) $\dim(M \cap W) = m - 1$, $\dim(N \cap W) = m + 1$.

The associated line of the spine space is said to be an *x-line*, $x \in \{\tau, \alpha, \omega\}$, if the corresponding case is realized. For example, the geometry of rectangular matrices ($m = 0$ and $\dim W = n - k$) contains only τ-lines.

Every automorphism of the spine space can be extended to an automorphism of the Grassmann space $\mathfrak{G}_k(V)$ [Prażmowski and Żynel (2002)].

Let B be a base of V such that W is spanned by a subset of B. The intersection of the associated apartment of $\mathcal{G}_k(V)$ with $\mathcal{F}_{k,m}(W)$ is said to be the *base subset* of $\mathcal{F}_{k,m}(W)$ defined by (associated with) the base B.

Two distinct elements of $\mathcal{F}_{k,m}(W)$ are called *x-adjacent*, $x \in \{\tau, \alpha, \omega\}$, if they belong to a certain x-line. Two x-adjacent elements of $\mathcal{F}_{k,m}(W)$ are contained in a base subset if and only if $x \in \{\alpha, \omega\}$.

In almost all cases, every bijective transformation of $\mathcal{F}_{k,m}(W)$ preserving the class of base subsets preserves the α-adjacency and ω-adjacency relations or interchanges them [Pankov, Prażmowski and Żynel (2005)]. The proof of this statement is based on a characterization of the α-adjacency and ω-adjacency relations in terms of complement subsets (the definitions of exact, inexact, and complement subsets are standard).

3.7 Geometry of linear involutions

3.7.1 *Involutions and transvections*

Suppose that V is an n-dimensional left vector space over a division ring R, the characteristic of R is not equal to 2 and $n \geq 3$.

Let u be an involution of the group $\mathrm{GL}(V)$ ($u^2 = 1$). Consider the linear subspaces

$$S_+(u) := \mathrm{Ker}\,(1 - u) \quad \text{and} \quad S_-(u) := \mathrm{Im}\,(1 - u).$$

We have

$$u(x) = x \ \text{ if } \ x \in S_+(u) \ \text{ and } \ u(x) = -x \ \text{ if } \ x \in S_-(u).$$

Since the intersection of these linear subspaces is 0 and the sum of their dimensions is n, we get

$$S_+(u) + S_-(u) = V.$$

We say that u is a $(k, n-k)$-*involution* if the dimension of $S_+(u)$ and $S_-(u)$ is equal to k and $n - k$, respectively.

Exercise 3.2. Let u be an involution of $\mathrm{GL}(V)$ and x be a vector satisfying $u(x) \in \langle x \rangle$. Show that x belongs to $S_+(u)$ or $S_-(u)$. *Hint:* there exist unique vectors $x_+ \in S_+(u)$ and $x_- \in S_-(u)$ such that $x = x_+ + x_-$.

The set of all $(k, n - k)$-involutions will be denoted by $\mathcal{I}_{k,n-k}(V)$. This set can be identified with

$$\mathcal{G}_{k,n-k}(V) := \{\, (S, U) \in \mathcal{G}_k(V) \times \mathcal{G}_{n-k}(V) \ : \ S + U = V \,\}.$$

It is clear that $v \in \mathrm{GL}(V)$ is conjugate to a $(k, n-k)$-involution u (there exists $l \in \mathrm{GL}(V)$ such that $v = lul^{-1}$) if and only if v is a $(k, n-k)$-involution. Therefore, each $\mathcal{I}_{k,n-k}(V)$ is a maximal set of conjugate involutions.

Lemma 3.8. *Involutions u and v commute if and only if*

$$u(S_+(v)) = S_+(v) \quad \text{and} \quad u(S_-(v)) = S_-(v).$$

Proof. Suppose that $uv = vu$. If $x \in S_+(v)$ then

$$vu(x) = uv(x) = u(x)$$

and $u(x) \in S_+(v)$. Similarly, we show that the linear subspace $S_-(v)$ is invariant for u.

Conversely, suppose that the linear subspaces $S_+(v)$ and $S_-(v)$ both are invariant for u. Then $uv(x) = vu(x)$ for every x belonging to $S_+(v)$ or $S_-(v)$. Since V is the sum of these linear subspaces, the latter equality holds for all $x \in V$. □

Let B be a base of V. The *base subset* of $\mathcal{I}_{k,n-k}(V)$ associated with (defined by) B is formed by all $(k, n-k)$-involutions u such that $S_+(u)$ and $S_-(u)$ both are spanned by subsets of the base B. By Lemma 3.8, any two elements of this subset commute.

Proposition 3.8. *The class of maximal commutative subsets of $\mathcal{I}_{k,n-k}(V)$ coincides with the class of base subsets.*

Proof. Let u and v be commutative involutions. Then $v(S_+(u)) = S_+(u)$ and the restriction of v to the linear subspace $S_+(u)$ is an involution; denote this involution by v'. It is clear that

$$S_+(v') = S_+(v) \cap S_+(u) \text{ and } S_-(v') = S_-(v) \cap S_+(u);$$

hence

$$S_+(v) \cap S_+(u) + S_-(v) \cap S_+(u) = S_+(u).$$

Similarly, we get the equality

$$S_+(v) \cap S_-(u) + S_-(v) \cap S_-(u) = S_-(u)$$

which implies that

$$S_+(v) \cap S_+(u) + S_-(v) \cap S_+(u) + S_+(v) \cap S_-(u) + S_-(v) \cap S_-(u) = V.$$

The same arguments show that for every commutative subset $\mathcal{X}$ of $\mathcal{I}_{k,n-k}(V)$ there exists a collection of linear subspaces $S_1, \ldots, S_m$ satisfying the following conditions:

- $S_1 + \cdots + S_m = V$,
- $S_i \cap S_j = 0$ if $i \neq j$,
- for every $u \in \mathcal{X}$ the linear subspaces $S_+(u)$ and $S_-(u)$ can be presented as the sums of some S_i.

This means that every commutative subset of $\mathcal{I}_{k,n-k}(V)$ is contained in a certain base subset and every base subset is a maximal commutative subset of $\mathcal{I}_{k,n-k}(V)$. $\qquad\square$

Let $t \in \mathrm{GL}(V)$. Suppose that $\mathrm{Ker}\,(1-t)$ is $(n-1)$-dimensional. Then the dimension of $\mathrm{Im}\,(1-t)$ is equal to 1 and

$$t(x) = x + \alpha(x)x_0,$$

where x_0 is a non-zero vector and α is a linear functional whose kernel coincides with $\mathrm{Ker}\,(1-t)$. If

$$\mathrm{Im}\,(1-t) \subset \mathrm{Ker}\,(1-t)$$

(in other words, $x_0 \in \mathrm{Ker}\,\alpha$) then t is said to be a *transvection*. The subgroup of $\mathrm{GL}(V)$ generated by all transvections is denoted by $\mathrm{SL}(V)$.

Remark 3.17. Note that $\alpha(x_0) \neq -1$ (since $t(x_0) \neq 0$) and

$$t^{-1}(x) = x - \frac{\alpha(x)}{1+\alpha(x_0)}x_0.$$

If t is a transvection then $t^{-1}(x) = x - \alpha(x)x_0$ also is a transvection.

Remark 3.18. If $l \in \mathrm{GL}(V)$ and t commute then

$$l(\mathrm{Ker}\,(1-t)) = \mathrm{Ker}\,(1-t) \quad \text{and} \quad l(\mathrm{Im}\,(1-t)) = \mathrm{Im}\,(1-t);$$

but these equalities do not guarantee that $lt = tl$.

3.7.2 *Adjacency relation*

Two distinct elements

$$(S,U), (S',U') \in \mathcal{G}_{k,n-k}(V)$$

are said to be *adjacent* if one of the following possibilities is realized:

- $S = S'$ and U is adjacent with U' in $\mathcal{G}_{n-k}(V)$,
- $U = U'$ and S is adjacent with S' in $\mathcal{G}_k(V)$.

The *Grassmann graph* $\Gamma_{k,n-k}(V)$ is the graph whose vertex set is $\mathcal{G}_{k,n-k}(V)$ and whose edges are pairs of adjacent elements.

We use the following lemma to show that this graph is connected.

Lemma 3.9. *For any distinct $S, S' \in \mathcal{G}_k(V)$ there exists $U \in \mathcal{G}_{n-k}(V)$ which is a complement to both S and S'.*

Proof.   Consider an independent subset

$$X \cup \{x_1, \ldots, x_m, y_1, \ldots, y_m\}$$

such that

$$S \cap S' = \langle X \rangle,$$

$$S = \langle X, x_1, \ldots, x_m \rangle,$$

$$S' = \langle X, y_1, \ldots, y_m \rangle.$$

If T is a complement to $S + S'$ then

$$U := T + \langle x_1 + y_1, \ldots, x_m + y_m \rangle$$

is as required. $\square$

Proposition 3.9. *The graph* $\Gamma_{k,n-k}(V)$ *is connected.*

Proof.   Let (S, U) and (S', U') be distinct elements of $\mathcal{G}_{k,n-k}(V)$. First, we consider the case when $U = U'$ and prove induction by the distance

$$d(S, S') = k - \dim(S \cap S')$$

that the graph $\Gamma_{k,n-k}(V)$ contains a path connecting (S, U) with (S', U). This is trivial if $d(S, S') = 1$. Suppose that $d(S, S') \geq 2$. We take any $(k-1)$-dimensional linear subspace W satisfying

$$S \cap S' \subset W \subset S.$$

Then $W + U \neq V$ and S' is not contained in $W + U$. This implies the existence of a 1-dimensional linear subspace $P' \subset S'$ such that $S'' := W + P'$ is a complement of U. We have

$$d(S', S'') = d(S, S') - 1.$$

By the inductive hypothesis, there exists a path in $\Gamma_{k,n-k}(V)$ connecting (S', U) with (S'', U). Since (S'', U) is adjacent with (S, U), we get the claim.

The case when $S = S'$ is similar.

In the general case, we take any $(n-k)$-dimensional linear subspace U'' satisfying

$$S + U'' = S' + U'' = V$$

(Lemma 3.9) and construct a path joining

$$(S, U), \ (S, U''), (S', U''), (S', U').$$ $\square$

For linear subspaces M and N we denote by $[M, N]_{k, n-k}$ the set of all $(S, U) \in \mathcal{G}_{k, n-k}(V)$ such that S and U are incident to M and N, respectively.

Proposition 3.10. *Every maximal clique of* $\Gamma_{1, n-1}(V)$ *is of type*

$$[P, V]_{1, n-1}, \ P \in \mathcal{G}_1(V) \quad or \quad [V, H]_{1, n-1}, \ H \in \mathcal{G}_{n-1}(V).$$

Similarly, every maximal clique of $\Gamma_{n-1, 1}(V)$ *is of type*

$$[V, P]_{n-1, 1}, \ P \in \mathcal{G}_1(V) \quad or \quad [H, V]_{n-1, 1}, \ H \in \mathcal{G}_{n-1}(V).$$

In the case when $1 < k < n - 1$, *the Grassmann graph* $\Gamma_{k, n-k}(V)$ *admits only the following four types of maximal cliques:*

- $[S, M]_{k, n-k}, \ S \in \mathcal{G}_k(V), \ M \in \mathcal{G}_{n-k-1}(V) \ and \ S \cap M = 0;$
- $[S, N]_{k, n-k}, \ S \in \mathcal{G}_k(V), \ N \in \mathcal{G}_{n-k+1}(V) \ and \ S + N = V;$
- $[Q, U]_{k, n-k}, \ Q \in \mathcal{G}_{k-1}(V), \ U \in \mathcal{G}_{n-k}(V) \ and \ Q \cap U = 0;$
- $[T, U]_{k, n-k}, \ T \in \mathcal{G}_{k+1}(V), \ U \in \mathcal{G}_{n-k}(V) \ and \ T + U = V.$

Proof. The case when $k = 1, n - 1$ is trivial. If $1 < k < n - 1$ then the statement easy follows from Proposition 3.2. $\qquad\square$

We say that two $(k, n - k)$-involutions are *adjacent* if the corresponding elements of $\mathcal{G}_{k, n-k}(V)$ are adjacent. There is the following algebraic interpretation of this adjacency relation.

Proposition 3.11. *For involutions* $u, v \in \mathcal{I}_{k, n-k}(V)$ *the following conditions are equivalent:*

(1) *u and v are adjacent,*
(2) *uv and vu are transvections.*

Proof. (1) $\implies$ (2). Suppose that $S_+(u) = S_+(v)$ (the case when $S_-(u) = S_-(v)$ is similar). Then

$$\mathrm{Ker}\,(1 - uv) = \mathrm{Ker}\,(1 - vu) = S_+(u) + (S_-(u) \cap S_-(v)) \qquad (3.6)$$

is $(n-1)$-dimensional. There exist a linear functional α whose kernel coincides with (3.6) and a non-zero vector x_0 such that

$$uv(x) = x + \alpha(x)x_0. \qquad (3.7)$$

Then

$$v(x) = u(x) + \alpha(x)u(x_0) \qquad (3.8)$$

and

$$x = v^2(x) = u(u(x) + \alpha(x)u(x_0)) + \alpha(u(x) + \alpha(x)u(x_0))u(x_0)$$

$$= x + \alpha(x)x_0 + \alpha(u(x) + \alpha(x)u(x_0))u(x_0)$$

which means that $u(x_0) \in \langle x_0 \rangle$. Then (3.8) implies that $v(x_0) \in \langle x_0 \rangle$. Using Exercise 3.2 we show that x_0 belongs to the linear subspace (3.6) which coincides with $\operatorname{Ker}\alpha$. Therefore, uv is a transvection; then vu also is a transvection (Remark 3.17).

(2) $\implies$ (1). If uv is a transvection then (3.7) holds for a certain linear functional α and a non-zero vector $x_0 \in \operatorname{Ker}\alpha$. As above, we establish that $u(x_0) \in \langle x_0 \rangle$. By Exercise 3.2, x_0 belongs to $S_+(u)$ or $S_-(u)$.

Suppose that $x_0 \in S_+(u)$. For every $x \in S_+(v)$

$$u(x) = uv(x) = x + \alpha(x)x_0$$

and

$$x = u^2(x) = u(x) + \alpha(x)u(x_0) = x + 2\alpha(x)x_0.$$

This implies that $S_+(v) \subset \operatorname{Ker}\alpha$. Since $u(x) = v(x)$ for every $x \in \operatorname{Ker}\alpha$, we have $S_+(v) = S_+(u)$ and

$$S_-(v) \cap (\operatorname{Ker}\alpha) = S_-(u) \cap (\operatorname{Ker}\alpha)$$

is $(n - k - 1)$-dimensional.

Suppose that $x_0 \in S_-(u)$. For every $x \in S_-(v)$

$$u(x) = -uv(x) = -x - \alpha(x)x_0.$$

As in the previous case, we get $S_-(v) \subset \operatorname{Ker}\alpha$ which implies that $S_-(v)$ coincides with $S_-(u)$ and $S_+(v)$ is adjacent with $S_+(u)$. $\qquad\square$

Let $\mathcal{X}$ be a non-empty subset of $\mathcal{I}_{k,n-k}(V)$. Denote by $\mathcal{X}^c$ the set consisting of all $(k, n - k)$-involutions commuting with every element of $\mathcal{X}$. If $\mathcal{X}^c$ is non-empty then $\mathcal{X}^{cc}$ contains $\mathcal{X}$.

Lemma 3.10 ([Mackey (1942)]). *Suppose that $k = 1, n - 1$. Then two distinct $(k, n-k)$-involutions u, v are adjacent if and only if for any distinct $u', v' \in \{u, v\}^{cc}$ we have*

$$\{u, v\}^{cc} = \{u', v'\}^{cc}. \tag{3.9}$$

Proof.   We restrict ourselves to the case when $k = 1$ (the case $k = n - 1$ is similar). Recall that every $(k, n - k)$-involution s is identified with

$$(S_+(s), S_-(s)) \in \mathcal{G}_{k,n-k}(V).$$

Let $u = (P, U)$ and $v = (P', U')$. By Lemma 3.8, a $(1, n - 1)$-involution (P'', U'') commute with u if and only if

$$P \subset U'' \quad \text{and} \quad P'' \subset U.$$

This implies that

$$\{u, v\}^c = [U \cap U', P + P']_{1,n-1}$$

and

$$\{u, v\}^{cc} = [P + P', U \cap U']_{1,n-1}.$$

If u and v are adjacent ($P = P'$ or $U = U'$) then any distinct $u', v' \in \{u, v\}^{cc}$ are adjacent and (3.9) holds. In the case when $P \neq P'$ and $U \neq U'$, the equality (3.9) fails for any adjacent $u', v' \in \{u, v\}^{cc}$. $\square$

Problem 3.1. Is there a characterization of the adjacency relation in terms of the commutativity relation for the case when $1 < k < n - 1$?

3.7.3   *Chow's theorem for linear involutions*

We give a few examples.

Example 3.10. For every semilinear automorphism $l : V \to V$ the mapping

$$u \to lul^{-1} \quad \forall\, u \in \mathrm{GL}(V)$$

is an automorphism of the group $\mathrm{GL}(V)$. It transfers each $\mathcal{I}_{k,n-k}(V)$ to itself. The associated transformation of $\mathcal{G}_{k,n-k}(V)$ is

$$(S, U) \to (l(S), l(U)).$$

Example 3.11. Recall that the contragradient mapping $u \to \check{u}$ is an isomorphism of $\mathrm{GL}(V)$ to $\mathrm{GL}(V^*)$. If u is a $(k, n - k)$-involution then the same holds for $\check{u}$; moreover,

$$S_+(\check{u}) = (S_-(u))^0 \quad \text{and} \quad S_-(\check{u}) = (S_+(u))^0$$

(an exercise for the reader). For every semilinear isomorphism $s : V \to V^*$ the mapping

$$u \to s^{-1}\check{u}s \quad \forall\, u \in \mathrm{GL}(V)$$

is an automorphism of the group $\mathrm{GL}(V)$. It transfers each $\mathcal{I}_{k,n-k}(V)$ to itself. The associated transformation of $\mathcal{G}_{k,n-k}(V)$ is

$$(S, U) \to (s^{-1}(U^0), s^{-1}(S^0)).$$

Automorphisms of the group $GL(V)$ considered in Examples 3.10 and 3.11 will be called *regular*.

Example 3.12. In the case when $n = 2k$, the mapping $u \to -u$ is a bijective transformation of $\mathcal{I}_{k,k}(V)$. The associated transformation of $\mathcal{G}_{k,k}(V)$ sends (S, U) to (U, S).

A bijective transformation f of $\mathcal{I}_{k,n-k}(V)$ is called *commutativity preserving* if f and f^{-1} map commutative involutions to commutative involutions; by Proposition 3.8, this is equivalent to the fact that f and f^{-1} transfer base subsets to base subsets.

It is clear that all transformations considered above are commutativity preserving and the associated transformations of $\mathcal{G}_{k,n-k}(V)$ are automorphisms of the Grassmann graph $\Gamma_{k,n-k}(V)$. In the case when $n = 2k$, there exist commutativity preserving transformations of $\mathcal{I}_{k,k}(V)$ which do not induce automorphisms of $\Gamma_{k,k}(V)$.

Example 3.13. Let $n = 2k$. Consider any proper subset $\mathcal{X} \subset \mathcal{I}_{k,k}(V)$ satisfying the following condition:

$$u \in \mathcal{X} \implies -u \in \mathcal{X}.$$

The transformation

$$u \to \begin{cases} -u & \forall \, u \in \mathcal{X} \\ u & \forall \, u \notin \mathcal{X} \end{cases}$$

is commutativity preserving; but the associated transformation of $\mathcal{G}_{k,k}(V)$ is not an automorphism of $\Gamma_{k,k}(V)$ (we leave the details for the reader).

Theorem 3.15 ([Havlicek and Pankov (2005)]). *If $n \neq 2k$ then every automorphism of $\Gamma_{k,n-k}(V)$ is induced by a semilinear automorphism of V (Example 3.10) or a semilinear isomorphism of V to V^* (Example 3.11). Let $n = 2k$ and f be an automorphism of $\Gamma_{k,k}(V)$. Then*

$$f = ig,$$

where g is the automorphism of $\Gamma_{k,k}(V)$ induced by a semilinear automorphism of V or a semilinear isomorphism of V to V^* and i is the transformation sending every (S, U) to (U, S).

Remark 3.19. Some generalizations of this result were obtained in [Prażmowski and Żynel (2009)].

Remark 3.20. It must be pointed out that all results concerning the adjacency relation on $\mathcal{G}_{k,n-k}(V)$ (Propositions 3.9, 3.10 and Theorem 3.15) hold in the case of an arbitrary characteristic.

Corollary 3.4 ([Dieudonné 1 (1951); Rickart (1950)]). *In the case when $k = 1, n - 1$, every commutativity preserving bijective transformation of $\mathcal{I}_{k,n-k}(V)$ can be extended to a regular automorphism of the group* GL(V).

Proof. Let $k = 1, n - 1$ and f be a commutativity preserving bijective transformation of $\mathcal{I}_{k,n-k}(V)$. By Lemma 3.10, the associated transformation of $\mathcal{G}_{k,n-k}(V)$ is an automorphism of $\Gamma_{k,n-k}(V)$. Theorem 3.15 gives the claim. □

Remark 3.21. Commutativity preserving transformations of $\mathcal{I}_{k,n-k}(V)$, $1 < k < n - 1$, were investigated in [Pankov (2005)].

3.7.4 *Proof of Theorem 3.15*

Let f be an automorphism of the graph $\Gamma_{k,n-k}(V)$. In what follows the subsets

$$[S, V]_{k,n-k},\ S \in \mathcal{G}_k(V)\ \text{ and }\ [V, U]_{k,n-k},\ U \in \mathcal{G}_{n-k}(V)$$

will be called *special*. Our first step is to show that f and f^{-1} map special subsets to special subsets. In the cases when $k = 1, n-1$, the class of special subsets coincides with the class of maximal cliques of $\Gamma_{k,n-k}(V)$ and the statement is trivial.

Suppose that $1 < k < n - 1$. Let $S \in \mathcal{G}_k(V)$. Consider $T \in \mathcal{G}_{n-k+1}(V)$ satisfying $S + T = V$. Then $[S, T]_{k,n-k}$ is a maximal clique of $\Gamma_{k,n-k}(V)$ and the same holds for $f([S, T]_{k,n-k})$. By Proposition 3.10, one of the four possibilities is realized. Assume that our cliques are of the same type; in other words, there exist

$$W \in \mathcal{G}_k(V)\ \text{ and }\ Z \in \mathcal{G}_{n-k+1}(V)$$

such that $W + Z = V$ and

$$f([S, T]_{k,n-k}) = [W, Z]_{k,n-k}.$$

Now we show that

$$f((S, U)) \in [W, V]_{k,n-k}\quad \forall\, (S, U) \in [S, V]_{k,n-k}.$$

Let us take any

$$(S, U_0) \in [S, T]_{k, n-k}$$

and suppose that (S, U) is adjacent with (S, U_0). Then the linear subspace $P := U \cap U_0$ is $(n - k - 1)$-dimensional and $[P, T]_{n-k}$ is a line of the Grassmann space $\mathfrak{G}_{n-k}(V)$. This line contains at least three elements and only one of them has a non-zero intersection with S. Thus the intersection of $[S, T]_{k, n-k}$ and $[S, P]_{k, n-k}$ contains more than one element and the same holds for their images

$$f([S, T]_{k, n-k}) = [W, Z]_{k, n-k} \text{ and } f([S, P]_{k, n-k}).$$

Since $f([S, P]_{k, n-k})$ is a maximal clique of $\Gamma_{k, n-k}(V)$, the linear subspace W is the first component in every element of $f([S, P]_{k, n-k})$ and

$$f((S, U)) \in [W, V]_{k, n-k}.$$

By the connectedness of the graph $\Gamma_{k, n-k}(V)$, the same holds for every $(S, U) \in [S, V]_{k, n-k}$. Therefore,

$$f([S, V]_{k, n-k}) \subset [W, V]_{k, n-k}.$$

We apply the same arguments to the mapping f^{-1} and get the inverse inclusion. In the case when $f([S, T]_{k, n-k})$ is a maximal clique of one of the remaining three types, the proof is similar.

Similarly, we show that for every $U \in \mathcal{G}_{n-k}(V)$ the image of the special subset $[V, U]_{k, n-k}$ is a special subset.

The intersection of two distinct special subsets is non-empty if and only if these special subsets are of different types and the sum of the associated k-dimensional and $(n-k)$-dimensional linear subspaces coincides with V (it is clear that this intersection is a one-element set). Since for any distinct $S_i \in \mathcal{G}_k(V)$ $(i = 1, 2)$ there exists $U \in \mathcal{G}_{n-k}(V)$ such that $S_i + U = V$ (Lemma 3.9), one of the following possibilities is realized:

(A) f preserves the type of every special subset,
(B) f changes the types of all special subsets.

Case (A). There exist bijective transformations f' and f'' of $\mathcal{G}_k(V)$ and $\mathcal{G}_{n-k}(V)$ (respectively) such that

$$f((S, U)) = (f'(S), f''(U)) \quad \forall (S, U) \in \mathcal{G}_{k, n-k}(V).$$

Thus for any $S \in \mathcal{G}_k(V)$ and $U \in \mathcal{G}_{n-k}(V)$

$$S + U = V \iff f'(S) + f''(U) = V.$$

The mappings f' and f'' are automorphisms of the Grassmann graphs $\Gamma_k(V)$ and $\Gamma_{n-k}(V)$, respectively (this statement is trivial if $k = 1, n - 1$, and it follows immediately from Theorem 3.7 if $1 < k < n - 1$).

Suppose that $k = 1, n - 1$. In this case, $S \in \mathcal{G}_k(V)$ and $U \in \mathcal{G}_{n-k}(V)$ are incident if and only if $f'(S)$ and $f''(U)$ are incident. By Proposition 3.4, f' and f'' are induced by the same semilinear automorphism of V.

If $1 < k < n - 1$ then each of these mappings is induced by a semilinear automorphism of V or a semilinear isomorphism of V to V^* (the second possibility is realized only in the case when $n = 2k$). Suppose that f' is induced by a semilinear automorphism $l : V \to V$. If f'' is induced by a semilinear automorphism $s : V \to V$ then for any $S \in \mathcal{G}_k(V)$ and $U \in \mathcal{G}_{n-k}(V)$

$$S + U = V \iff S + l^{-1}s(U) = V;$$

this implies that $l^{-1}(s(U)) = U$ for every $U \in \mathcal{G}_{n-k}(V)$ and

$$f((S, U)) = (l(S), l(U)) \quad \forall\, (S, U) \in \mathcal{G}_{k, n-k}(V).$$

Now, suppose that $n = 2k$ and f'' is induced by a semilinear isomorphism $s : V \to V^*$. As above, we establish that

$$(sl^{-1}(U))^0 = U \quad \forall\, U \in \mathcal{G}_k(V).$$

Let $W \in \mathcal{G}_{k-1}(V)$. We choose

$$U_1, \ldots, U_{k+1} \in \mathcal{G}_k(V)$$

such that

$$U_1 \cap \cdots \cap U_{k+1} = W \quad \text{and} \quad U_1 + \cdots + U_{k+1} = V.$$

Then

$$0 = (sl^{-1}(V))^0 = \bigcap_{i=1}^{k+1} (sl^{-1}(U_i))^0 = \bigcap_{i=1}^{k+1} U_i = W.$$

The equality $W = 0$ implies that $k = 1$ which contradicts the assumption that $n = 2k \geq 3$.

If $n = 2k$ and f' is induced by a semilinear isomorphism $u : V \to V^*$ then the same arguments show that f'' also is induced by u; thus f is the composition of the transformations

$$(S, U) \to (u(U)^0, u(S)^0) \quad \text{and} \quad (S, U) \to (U, S).$$

Case (B). There exist bijective mappings

$$g' : \mathcal{G}_k(V) \to \mathcal{G}_{n-k}(V) \quad \text{and} \quad g'' : \mathcal{G}_{n-k}(V) \to \mathcal{G}_k(V)$$

such that

$$f((S,U)) = (g''(U), g'(S)) \quad \forall\, (S,U) \in \mathcal{G}_{k,n-k}(V).$$

By duality, these mappings can be considered as bijections of $\mathcal{G}_k(V)$ and $\mathcal{G}_{n-k}(V)$ to $\mathcal{G}_k(V^*)$ and $\mathcal{G}_{n-k}(V^*)$, respectively. As in the case (A), we establish that one of the following possibilities is realized: (i) g' and g'' are induced by the same similinear isomorphism $u : V \to V^*$ and

$$f((S,U)) = (u(U)^0, u(S)^0) \quad \forall\, (S,U) \in \mathcal{G}_{k,n-k}(V);$$

(ii) $n = 2k$, the mappings g' and g'' are induced by the same similinear automorphism $l : V \to V'$ and f is the composition of the transformations

$$(S,U) \to (l(S), l(U)) \quad \text{and} \quad (S,U) \to (U,S).$$

3.7.5   *Automorphisms of the group* $\mathrm{GL}(V)$

Theorem 3.16 ([Dieudonné 1 (1951); Rickart (1950)]). *If f is an automorphism of the group $\mathrm{GL}(V)$ then*

$$f(u) = \alpha(u) g(u) \quad \forall\, u \in \mathrm{GL}(V),$$

where g is a regular automorphism and α is a homomorphism of $\mathrm{GL}(V)$ to the center of R.

Proof.   The automorphism f preserves the set of all involutions; moreover, it transfers $\mathcal{I}_{k,n-k}(V)$ to $\mathcal{I}_{m,n-m}(V)$ (because each $\mathcal{I}_{k,n-k}(V)$ can be characterized as a maximal set of conjugate involutions). By Proposition 3.8, base subsets of $\mathcal{I}_{k,n-k}(V)$ go to base subsets of $\mathcal{I}_{m,n-m}(V)$. Since a base subset of $\mathcal{I}_{k,n-k}(V)$ consists of $\binom{n}{k}$ elements and

$$\binom{n}{k} = \binom{n}{m} \iff m = k, n-k,$$

the image of $\mathcal{I}_{k,n-k}(V)$ coincides with $\mathcal{I}_{k,n-k}(V)$ or $\mathcal{I}_{n-k,k}(V)$. Therefore, the restriction of f to the set $\mathcal{I}_{1,n-1}(V)$ is a commutativity preserving bijection to $\mathcal{I}_{1,n-1}(V)$ or $\mathcal{I}_{n-1,1}(V)$. By Corollary 3.4, there exists a regular automorphism g of $\mathrm{GL}(V)$ such that

$$f(u) = g(u) \quad \forall\, u \in \mathcal{I}_{1,n-1}(V) \quad \text{or} \quad f(u) = -g(u) \quad \forall\, u \in \mathcal{I}_{1,n-1}(V).$$

For every transvection t we take adjacent $(1, n-1)$-involutions u, v satisfying $t = uv$ and get $f(t) = g(t)$ in each of these cases. So, f and g are coincident on $\mathrm{SL}(V)$ and $h := g^{-1}f$ sends every element of $\mathrm{SL}(V)$ to itself.

If t is a transvection then ltl^{-1} is a transvection for any $l \in \mathrm{GL}(V)$ and

$$h(l)th(l^{-1}) = ltl^{-1}$$

which means that $\alpha(l) := l^{-1}h(l)$ commutes with every transvection. By Remark 3.18, $\alpha(l)$ preserves every 1-dimensional linear subspace; hence it is a homothetic transformation. All linear homothetic transformations of V form the center of the group $\mathrm{GL}(V)$ (this subgroup is isomorphic to the center of R). The mapping $l \to \alpha(l)$ is a homomorphism of $\mathrm{GL}(V)$ to the center. An easy verification shows that $f(l) = \alpha(l)g(l)$. $\square$

3.8 Grassmannians of infinite-dimensional vector spaces

Let V and V' be vector spaces over division rings. Throughout the section we suppose that $\dim V = \dim V' = \alpha$ is infinite. In this case, we have the following three types of Grassmannians associated with V:

$$\mathcal{G}_\beta(V) = \{ \, S \in \mathcal{G}(V) \ : \ \dim S = \beta, \ \mathrm{codim}\, S = \alpha \, \},$$

$$\mathcal{G}^\beta(V) = \{ \, S \in \mathcal{G}(V) \ : \ \dim S = \alpha, \ \mathrm{codim}\, S = \beta \, \}$$

for every cardinality $\beta < \alpha$ and

$$\mathcal{G}_\alpha(V) = \mathcal{G}^\alpha(V) = \{ \, S \in \mathcal{G}(V) \ : \ \dim S = \mathrm{codim}\, S = \alpha \, \}.$$

They are orbits of the action of the group $\mathrm{GL}(V)$ on the set $\mathcal{G}(V)$.

3.8.1 *Adjacency relation*

Let $\mathcal{G}$ be a Grassmannian of V. We say that linear subspaces $S, U \in \mathcal{G}$ are *adjacent* if

$$\dim(S/(S \cap U)) = \dim(U/(S \cap U)) = 1$$

which is equivalent to

$$\dim((S + U)/S) = \dim((S + U)/U) = 1.$$

We define the associated Grassmann graph as in the finite-dimensional case. The Grassmann graphs corresponding to $\mathcal{G}_\beta(V)$, $\mathcal{G}^\beta(V)$, and $\mathcal{G}_\alpha(V)$ will be denoted by $\Gamma_\beta(V)$, $\Gamma^\beta(V)$, and $\Gamma_\alpha(V)$ (respectively). Semilinear isomorphisms of V to V' induce isomorphisms between the Grassmann graphs of the same indices.

Let $k \in \mathbb{N}$. The Grassmann graph $\Gamma_k(V)$ is connected and every isomorphism of $\Gamma_k(V)$ to $\Gamma_k(V')$ is induced by a semilinear isomorphism of

V to V' (the proof is similar to the proof of Theorem 3.2 and we leave it as an exercise for the reader). Since the annihilator mapping defines an isomorphism of $\Gamma^k(V)$ to $\Gamma_k(V^*)$ (see Subsection 1.1.3), the Grassmann graph $\Gamma^k(V)$ is connected.

The following example shows that the direct analogue of Chow's theorem does not hold for Grassmannians formed by linear subspaces of infinite dimension and codimension.

Example 3.14 ([Blunck and Havlicek (2005)]). Let $S \in \mathcal{G}_\beta(V)$ and β be infinite. Denote by $\mathcal{X}$ the connected component of $\Gamma_\beta(V)$ containing S; it consists of all $X \in \mathcal{G}_\beta(V)$ such that

$$\dim(S/(S \cap X)) = \dim(X/(S \cap X))$$

is finite. We take any $S' \in \mathcal{X} \setminus \{S\}$. There exists $U \in \mathcal{G}_\beta(V) \setminus \mathcal{X}$ such that

$$U \cap S = 0 \quad \text{and} \quad U \cap S' \neq 0.$$

Let $l : V \to V$ be a semilinear automorphism sending S to S'. The associated automorphism of $\Gamma_\beta(V)$ transfers $\mathcal{X}$ to itself. We define

$$f(X) := \begin{cases} l(X) & X \in \mathcal{X} \\ X & X \in \mathcal{G}_\beta(V) \setminus \mathcal{X}. \end{cases}$$

This mapping is an automorphism of $\Gamma_\beta(V)$, but it is not induced by a semilinear automorphism of V. Indeed, $U \cap S = 0$ and $f(U) = U$ has a non-zero intersection with $f(S) = S'$.

Problem 3.2. Let $\mathcal{X}$ be a connected component of $\Gamma_\beta(V)$, $\beta < \alpha$, and f be an automorphism of $\Gamma_\beta(V)$. Is there a semilinear automorphism $l : V \to V$ such that $f(S) = l(S)$ for all $S \in \mathcal{X}$?

There is the following infinite-dimensional version of Theorem 3.7.

Theorem 3.17 ([Blunck and Havlicek (2005)]). *Let $\mathcal{G}$ be a Grassmannian of V. For any distinct $S_1, S_2 \in \mathcal{G}$ the following conditions are equivalent:*

(1) *S_1 and S_2 are adjacent,*
(2) *there exists $S \in \mathcal{G} \setminus \{S_1, S_2\}$ such that every complement of S is a complement to at least one of S_i.*

For every $S \in \mathcal{G}_\alpha(V)$ all complements of S belong to $\mathcal{G}_\alpha(V)$ (for other Grassmannians this fails) and we have the following.

Corollary 3.5. *Suppose that $f : \mathcal{G}_\alpha(V) \to \mathcal{G}_\alpha(V')$ is a bijection preserving the complementary of subspaces: $S \in \mathcal{G}_\alpha(V)$ is a complement of $U \in \mathcal{G}_\alpha(V)$ if and only if $f(S)$ is a complement of $f(U)$. Then f is an isomorphism of $\Gamma_\alpha(V)$ to $\Gamma_\alpha(V')$.*

3.8.2 *Proof of Theorem 3.17*

If $\mathcal{G} = \mathcal{G}_1(V), \mathcal{G}^1(V)$ then any two distinct elements of $\mathcal{G}$ are adjacent and the implication (2) $\Longrightarrow$ (1) is trivial.

(1) $\Longrightarrow$ (2). If S_1 and S_2 are adjacent then any $S \in \mathcal{G} \setminus \{S_1, S_2\}$ satisfying

$$S_1 \cap S_2 \subset S \subset S_1 + S_2$$

is as required (see the proof of Theorem 3.6).

(2) $\Longrightarrow$ (1). Suppose that $S \in \mathcal{G} \setminus \{S_1, S_2\}$ satisfies the condition (2). As in the proof of Theorem 3.6, we establish that for any 1-dimensional linear subspaces $P_1 \subset S_1$, $P_2 \subset S_2$ the sum $P_1 + P_2$ has a non-zero intersection with S and

$$S_1 \cap S_2 \subset S. \tag{3.10}$$

Now we show that

$$S \subset S_1 + S_2. \tag{3.11}$$

This inclusion is trivial if $S_1 + S_2 = V$. In the case when $S_1 + S_2 \neq V$, it is sufficient to show that S is contained in every $H \in \mathcal{G}^1(V)$ containing $S_1 + S_2$ (since the intersection of all such linear subspaces coincides with $S_1 + S_2$). If $S \not\subset H \in \mathcal{G}^1(V)$ then $H + S = V$ and H contains a complement U of S; if $S_1 + S_2 \subset H$ then U is not a complement of S_i ($i = 1, 2$) which contradicts (2).

The linear subspaces S_1 and S_2 are not incident. Indeed, if $S_1 \subset S_2$ then (3.10) and (3.11) imply that

$$S_1 \subset S \subset S_2$$

and every complement of S is not a complement of S_i ($i = 1, 2$); similarly, we get $S_2 \not\subset S_1$.

Our next step is to show that

$$S_i \not\subset S, \quad i = 1, 2.$$

Suppose, for example, that $S_1 \subset S$. Since $S_1 \not\subset S_2$, there exists $H \in \mathcal{G}^1(V)$ which contains S_2 and does not contain S_1. We have

$$V = S_1 + H \subset S + H;$$

thus $S + H = V$ and H contains a certain complement U of S. Then U is not a complement of S_1 (S_1 is a proper subspace of S) and it is not a complement of S_2 (U and S_2 both are contained in H), a contradiction.

We take any 2-dimensional linear subspace $U \subset S_1$ and a 1-dimensional linear subspace of S_2 which is not contained in S. As in the proof of Theorem 3.6, we establish that U has a non-zero intersection with $S \cap S_1$. This means that

$$\dim(S_1/(S \cap S_1)) = 1$$

and the same holds for S_2. Thus there exist 1-dimensional linear subspaces $P_1 \subset S_1$ and $P_2 \subset S_2$ such that

$$S_i = (S \cap S_i) + P_i, \quad i = 1, 2.$$

Note that $P_1 \neq P_2$ (otherwise $P_1 = P_2 \subset S_1 \cap S_2 \subset S$ and both S_1, S_2 are contained in S).

The linear subspace $P_1 + P_2$ intersects S in a 1-dimensional linear subspace P. We have

$$S_1 + S_2 \subset S + P_1 + P_2 = S + P + P_1 = S + P_1 \subset S_1 + S_2$$

which implies that

$$S_1 + S_2 = S + P_1. \tag{3.12}$$

By the same arguments,

$$S_1 + S_2 = S + P_2. \tag{3.13}$$

Let U be a complement of $S_1 + S_2$. By (3.12), $U + P_1$ is a complement of S. Since $P_1 \subset S_1$, the linear subspace $U + P_1$ is not a complement of S_1; hence it is a complement of S_2. So,

$$U + (S_1 + S_2) = V = U + (S_2 + P_1)$$

and the inclusion $S_2 + P_1 \subset S_1 + S_2$ implies that

$$S_1 + S_2 = S_2 + P_1.$$

Using the equality (3.13), we establish that

$$S_1 + S_2 = S_1 + P_2.$$

Therefore,

$$\dim((S_1 + S_2)/S_1) = \dim((S_1 + S_2)/S_2) = 1$$

which means that S_1 and S_2 are adjacent.

3.8.3   Base subsets

Let $\mathcal{G}$ be a Grassmannian of V and B be a base of V. The *base subset* of $\mathcal{G}$ associated with (defined by) B consists of all elements of $\mathcal{G}$ spanned by subsets of B. For any two elements of $\mathcal{G}$ there exists a base subsets containing them (Proposition 1.4).

Every base subset of $\mathcal{G}_1(V)$ is a base of the projective space Π_V and conversely. Base subsets of $\mathcal{G}^1(V)$ are not bases of the dual projective space $\Pi_V^* = \Pi_{V^*}$, since $\dim V < \dim V^*$ (Subsection 1.1.3).

The mappings between Grassmannians of the same indices induced by semilinear isomorphisms of V to V' transfer base subsets to base subsets. We prove the following generalization of Theorem 3.9.

Theorem 3.18 ([Pankov 1 (2007)]). *Let f be a bijection of $\mathcal{G}_\beta(V)$ to $\mathcal{G}_\beta(V')$, $\beta < \alpha$, such that f and f^{-1} map base subsets to base subsets. Then f is induced by a semilinear isomorphism of V to V'.*

3.8.4   Proof of Theorem 3.18

If $\beta = 1$ then, as in Subsection 3.4.1, we show that f and f^{-1} are semi-collineations between the projective spaces associated with V and V'; thus f is a collineation of Π_V to $\Pi_{V'}$ and the Fundamental Theorem of Projective Geometry gives the claim.

Let $\mathcal{G}$ be a Grassmannian of V distinct from $\mathcal{G}_1(V)$ and $\mathcal{G}^1(V)$. Let also $B = \{x_i\}_{i \in I}$ be a base of V and $\mathcal{B}$ be the associated base subset of $\mathcal{G}$. *Exact, inexact,* and *complement* subsets of $\mathcal{B}$ are defined as for apartments in Grassmannians of finite-dimensional vector spaces. We write $\mathcal{B}(+i)$ and $\mathcal{B}(-i)$ for the sets of all elements of $\mathcal{B}$ which contain x_i and do not contain x_i, respectively. The subsets $\mathcal{B}(+i)$ and $\mathcal{B}(-i)$ will be called *simple subsets* of *first* and *second type*, respectively. We also define

$$\mathcal{B}(+i, +j) := \mathcal{B}(+i) \cap \mathcal{B}(+j) \text{ and } \mathcal{B}(+i, -j) := \mathcal{B}(+i) \cap \mathcal{B}(-j)$$

for all $i, j \in I$.

As in Subsection 3.4.2, we establish that for every maximal inexact subset $\mathcal{X} \subset \mathcal{B}$ there exist distinct $i, j \in I$ such that

$$\mathcal{X} = \mathcal{B}(+i, +j) \cup \mathcal{B}(-i).$$

The associated complement subset is $\mathcal{B}(+i, -j)$. Note that every complement subset is the intersection of two simple subsets of different types.

For two distinct complement subsets $\mathcal{B}(+i, -j)$ and $\mathcal{B}(+i', -j')$ one of the following possibilities is realized:

(1) $i = i'$ or $j = j'$,

(2) $i = j'$ or $j = i'$, then the intersection of the complement subsets is empty,

(3) $\{i, j\} \cap \{i', j'\} = \emptyset$.

In the first case, our complement subsets are said to be *adjacent*. An easy verification shows that for two distinct complement subsets $\mathcal{X}, \mathcal{Y} \subset \mathcal{B}$ with a non-empty intersection the following two conditions are equivalent:

- $\mathcal{X}$ and $\mathcal{Y}$ are adjacent,
- for any distinct complement subsets $\mathcal{X}', \mathcal{Y}' \subset \mathcal{B}$ satisfying

$$\mathcal{X} \cap \mathcal{Y} \subset \mathcal{X}' \cap \mathcal{Y}'$$

the inverse inclusion holds.

In other words, the intersection of two distinct complement subsets is maximal if and only if these complement subsets are adjacent.

A collection of mutually adjacent complement subsets of $\mathcal{B}$ will be called an *A-collection*. For each $i \in I$

$$\{\mathcal{B}(+i, -j)\}_{j \in I \setminus \{i\}} \quad \text{and} \quad \{\mathcal{B}(+j, -i)\}_{j \in I \setminus \{i\}}$$

are maximal *A*-collections. It is easy to see that every maximal *A*-collection is a collection of such kind. Thus every simple subset of $\mathcal{B}$ can be characterized as the union of all complement subsets belonging to a certain maximal *A*-collection.

Let $\mathcal{B}'$ be the base subset of $\mathcal{G}$ associated with other base $B' = \{x_i'\}_{i \in I}$. The simple subsets of $\mathcal{B}'$ corresponding to $i \in I$ will be denoted by $\mathcal{B}'(+i)$ and $\mathcal{B}'(-i)$.

A bijection $g : \mathcal{B} \to \mathcal{B}'$ is said to be *special* if g and g^{-1} map inexact subsets to inexact subsets.

Lemma 3.11. *Let* $g : \mathcal{B} \to \mathcal{B}'$ *be a special bijection. Then* g *and* g^{-1} *send simple subsets to simple subsets; moreover, there exists a bijective transformation* $\delta : I \to I$ *such that*

$$g(\mathcal{B}(+i)) = \mathcal{B}'(+\delta(i)), \quad g(\mathcal{B}(-i)) = \mathcal{B}'(-\delta(i)) \qquad \forall \, i \in I$$

or

$$g(\mathcal{B}(+i)) = \mathcal{B}'(-\delta(i)), \quad g(\mathcal{B}(-i)) = \mathcal{B}'(+\delta(i)) \qquad \forall \, i \in I.$$

Proof. It is clear that maximal inexact subsets go to maximal inexact subsets in both directions. Thus g and g^{-1} map complement subsets to complement subsets. Since two distinct complement subsets are adjacent if and only if their intersection is maximal, the adjacency relations of complement subsets is preserved and maximal A-collections go to maximal A-collections (in both directions). Then g and g^{-1} transfer simple subsets to simple subsets. Two distinct simple subsets are of different types if and only if their intersection is empty or a complement subset. This means that g and g^{-1} map simple subsets of different types to simple subsets of different types; hence they preserve the types of all simple subsets or change the type of every simple subset. Since $\mathcal{B}(-i) = \mathcal{B} \setminus \mathcal{B}(+i)$, there exists a bijective transformation $\delta : I \to I$ satisfying the required condition. $\square$

Let $g : \mathcal{B} \to \mathcal{B}'$ be a special bijection. We say that g is a *special bijection of the first type* if it preserves the types of all simple subsets; otherwise, g is said to be a *special bijection of the second type*.

Let $S, U \in \mathcal{B}$. The equality

$$S \cap U = \langle x_i \rangle$$

implies that the dimension of S and U is not greater than the codimension; hence $\mathcal{G} = \mathcal{G}_\beta(V)$, $\beta \leq \alpha$. This equality is equivalent to the fact that $S, U \in \mathcal{B}(+i)$ and for every $j \in I \setminus \{i\}$ we have $S \notin \mathcal{B}(+j)$ or $U \notin \mathcal{B}(+j)$.

Similarly, the equality

$$S + U = \langle B \setminus \{x_i\} \rangle$$

implies that the dimension of S and U is α; then $\mathcal{G} = \mathcal{G}^\beta(V)$, $\beta \leq \alpha$. This equality is equivalent to the fact that $S, U \in \mathcal{B}(-i)$ and for every $j \in I \setminus \{i\}$ we have $S \notin \mathcal{B}(-j)$ or $U \notin \mathcal{B}(-j)$.

We have proved the following.

Lemma 3.12. *Let g and δ be as in the previous lemma. Let also $S, U \in \mathcal{B}$. If g is a special bijection of the first type and $\mathcal{G} = \mathcal{G}_\beta(V)$, $\beta \leq \alpha$, then*

$$S \cap U = \langle x_i \rangle \iff g(S) \cap g(U) = \langle x'_{\delta(i)} \rangle.$$

If g is a special bijection of the first type and $\mathcal{G} = \mathcal{G}^\beta(V)$, $\beta \leq \alpha$, then

$$S + U = \langle B \setminus \{x_i\} \rangle \iff g(S) + g(U) = \langle B' \setminus \{x'_{\delta(i)}\} \rangle.$$

If g is a special bijection of the second type then

$$S \cap U = \langle x_i \rangle \iff g(S) + g(U) = \langle B' \setminus \{x'_{\delta(i)}\} \rangle,$$

$$S + U = \langle B \setminus \{x_i\} \rangle \iff g(S) \cap g(U) = \langle x'_{\delta(i)} \rangle,$$

and $\mathcal{G} = \mathcal{G}_\alpha(V)$.

By Lemma 3.12, special bijections of the second type exist only in the case when $\mathcal{G} = \mathcal{G}_\alpha(V)$.

For every linear subspace $U \subset V$ we denote by $[U]$ the set of all elements of $\mathcal{G}$ incident with U.

Lemma 3.13. *Suppose that $\mathcal{G} = \mathcal{G}_\beta(V)$, $\beta \le \alpha$. Then for every $i \in I$ and every $S \in [\langle x_i \rangle] \setminus \mathcal{B}$ there exist $M, N \in \mathcal{B}$ such that*

$$M \cap N = \langle x_i \rangle \tag{3.14}$$

and M, N, S are contained in a base subset of $\mathcal{G}$.

Proof. Consider the set of all $X \subset B$ such that $S \cap \langle X \rangle = 0$. Using Zorn lemma, we establish the existence of a maximal subset $X \subset B$ satisfying this condition (see the proof of Proposition 1.3). Then $\langle X \rangle$ is a complement of S (for every vector $x_j \in B \setminus X$ the linear subspace $\langle X, x_j \rangle$ has a non-zero intersection with S and x_j belongs to $S + \langle X \rangle$). Therefore, the cardinality of X is equal to α. This implies the existence of linear subspaces M', N' spanned by non-intersecting subsets of X and such that

- $\dim M' = \dim N' = \beta$ if β is infinite,
- $\dim M' = \dim N' = \beta - 1$ if β is finite.

The linear subspaces

$$M := \langle M', x_i \rangle \quad \text{and} \quad N := \langle N', x_i \rangle$$

belong to $\mathcal{B}$ and satisfy (3.14). We take any base Y of S containing x_i. It is not difficult to prove that $X \cup Y$ is a base of V (see the proof of Proposition 1.4). The associated base subset of $\mathcal{G}$ contains M, N, S. □

Remark 3.22. In the case when $\mathcal{G} = \mathcal{G}^\beta(V)$, $\beta \le \alpha$, we need the following "dual" version of Lemma 3.13: if

$$S_i := \langle B \setminus \{x_i\} \rangle, \quad i \in I,$$

then for every $S \in [S_i] \setminus \mathcal{B}$ there exist $M, N \in \mathcal{B}$ such that

$$M + N = S_i$$

and M, N, S are contained in a base subset of $\mathcal{G}$. The dual principles do not work and this statement cannot be obtained immediately from Lemma 3.13.

Now we prove the theorem. Let $f : \mathcal{G}_\beta(V) \to \mathcal{G}_\beta(V')$, $\beta < \alpha$, be a bijection such that f and f^{-1} map base subsets to base subsets.

Let $P \in \mathcal{G}_1(V)$. We take any base of V which contains a vector belonging to P and consider the associated base subset $\mathcal{B} \subset \mathcal{G}_\beta(V)$. The restriction of f to $\mathcal{B}$ is a special bijection to the base subset $f(\mathcal{B})$. By Lemma 3.12, this is a special bijection of first type and there exists $h(P) \in \mathcal{G}_1(V')$ such that

$$f(\mathcal{B} \cap [P]) \subset [h(P)].$$

Let $S \in [P] \setminus \mathcal{B}$. Lemma 3.13 implies the existence of $M, N \in \mathcal{B} \cap [P]$ satisfying $M \cap N = P$ and a base subset $\hat{\mathcal{B}} \subset \mathcal{G}_\beta(V)$ containing M, N, S. We have

$$f(\hat{\mathcal{B}} \cap [P]) \subset [P']$$

for a certain $P' \in \mathcal{G}_1(V')$. Lemma 3.12 guarantees that

$$h(P) = f(M) \cap f(N) = P'$$

and $f(S)$ belongs to $[h(P)]$. So,

$$f([P]) \subset [h(P)].$$

We apply the same arguments to f^{-1} and establish the inverse inclusion.

Therefore, there exists a mapping $h : \mathcal{G}_1(V) \to \mathcal{G}_1(V')$ satisfying

$$f([P]) = [h(P)] \quad \forall \, P \in \mathcal{G}_1(V).$$

The mapping h is bijective (the inverse mapping is induced by f^{-1}) and the latter equality implies that

$$h(\mathcal{G}_1(S)) = \mathcal{G}_1(f(S)) \quad \forall \, S \in \mathcal{G}_\beta(V). \tag{3.15}$$

Let B be a base of V and $\mathcal{B}$ be the associated base subset of $\mathcal{G}_\beta(V)$. Let also B' be one of the bases of V' associated with the base subset $f(\mathcal{B})$. We write $\mathcal{B}_1$ and $\mathcal{B}'_1$ for the bases of Π_V and $\Pi_{V'}$ defined by B and B', respectively. It follows from Lemma 3.12 that $h(\mathcal{B}_1) = \mathcal{B}'_1$.

So, h and h^{-1} send projective bases to projective bases. This means that h is induced by a semilinear isomorphism $l : V \to V'$. Then

$$h(\mathcal{G}_1(S)) = \mathcal{G}_1(l(S))$$

for every linear subspace $S \subset V$ and, by (3.15), we have $f(S) = l(S)$ for every $S \in \mathcal{G}_\beta(V)$.

Chapter 4

Polar and Half-Spin Grassmannians

All thick buildings of types C_n ($n \geq 3$) and D_n ($n \geq 4$) can be obtained from *polar spaces* of rank n. By [Buekenhout and Shult (1974)], polar spaces can be defined as partial linear spaces satisfying some natural axioms (each line contains at least three points, there is no point collinear with all other points, ...) and Buekenhout–Shult's well-known property which says that a point is collinear with one or all points of a line. Using Teirlinck's characterization of projective spaces, we show that this definition is equivalent to the classical Tits–Veldkamp definition of polar spaces (Theorem 4.1); in particular, all maximal singular subspaces of a polar space are projective spaces of the same finite dimension m (the number $m + 1$ is called the rank of the polar space). One of our main objects is the *polar Grassmannian* $\mathcal{G}_k(\Pi)$, $k \in \{0, 1, \ldots, n - 1\}$, consisting of all k-dimensional singular subspaces of a rank n polar space Π. Basic properties of polar spaces and their Grassmannians will be studied in Sections 4.1 and 4.2. In particular, it will be shown that for every rank n polar space one of the following possibilities is realized:

- each $(n-2)$-dimensional singular subspace is contained in at least three distinct maximal singular subspaces (type C_n),
- each $(n-2)$-dimensional singular subspace is contained in precisely two maximal singular subspaces (type D_n).

Section 4.3 is dedicated to examples: polar spaces associated with reflexive sesquilinear forms and quadratic forms, polar spaces of type D_3 (these polar spaces are isomorphic to the index two Grassmann spaces of 4-dimensional vector spaces). We do not consider polar spaces obtained from pseudo-quadratic forms and polar spaces of type C_3 associated with Cayley algebras. Remarks concerning embeddings in projective spaces and classification of polar spaces finish the section.

In Sections 4.4 and 4.5 we consider polar buildings and investigate elementary properties of their Grassmann spaces. Let Π be a polar space of rank n and $\Delta(\Pi)$ be the simplicial complex consisting of all flags formed by singular subspaces of Π. Then $\Delta(\Pi)$ is a building of type C_n. The Grassmannians of this building are the polar Grassmannians of Π. The corresponding Grassmann spaces will be denoted by $\mathfrak{G}_k(\Pi)$. The building $\Delta(\Pi)$ is thick only in the case when Π is a polar space of type C_n. Suppose that Π is a polar space of type D_n. Then $\mathcal{G}_{n-1}(\Pi)$ can be naturally decomposed in two disjoint subsets called the *half-spin Grassmannians* and denoted by $\mathcal{G}_+(\Pi)$ and $\mathcal{G}_-(\Pi)$. As in Example 2.3, the polar space Π defines a thick building of type D_n. The Crassmannians of this building are the polar Grassmannians $\mathcal{G}_k(\Pi)$, $k \le n - 3$, and the half-spin Grassmannians. As in the previous case, the Grassmann space associated with $\mathcal{G}_k(\Pi)$ is $\mathfrak{G}_k(\Pi)$. The Grassmann spaces of the half-spin Grassmannians are denoted by $\mathfrak{G}_\delta(\Pi)$, $\delta \in \{+, -\}$.

Let Π and Π' be polar spaces of the same type X_n, $\mathsf{X} \in \{\mathsf{C}, \mathsf{D}\}$, and $n \ge 3$; in the case when $\mathsf{X} = \mathsf{D}$, we require that $n \ge 4$. In Section 4.6 we investigate collineations of $\mathfrak{G}_k(\Pi)$ to $\mathfrak{G}_k(\Pi')$. We also consider collineations of $\mathfrak{G}_\delta(\Pi)$ to $\mathfrak{G}_\gamma(\Pi')$, $\delta, \gamma \in \{+, -\}$, if our polar spaces are of type D_n. In almost all cases, such collineations are induced by collineations of Π to Π'. However, if our polar spaces are of type D_4 then their half-spin Grassmann spaces (the Grassmann spaces of the half-spin Grassmannians) are polar spaces of type D_4 and there are two additional possibilities:

- the collineations of $\mathfrak{G}_1(\Pi)$ to $\mathfrak{G}_1(\Pi')$ induced by collineations of Π to $\mathfrak{G}_\delta(\Pi')$, $\delta \in \{+, -\}$;
- the collineations of $\mathfrak{G}_\delta(\Pi)$ to $\mathfrak{G}_\gamma(\Pi')$, $\delta, \gamma \in \{+, -\}$, induced by collineations of Π to $\mathfrak{G}_{-\gamma}(\Pi')$.

In Section 4.7 we give an example showing that the adjacency relation on $\mathcal{G}_{n-1}(\Pi)$ cannot be characterized in terms of the opposite relation as in Theorem 3.6; however, the direct analogue of Theorem 3.6 holds for the half-spin Grassmannians. Characterizations of apartments in terms of the adjacency relation (analogues of Theorems 3.8) will be obtained only for the polar Grassmannians formed by maximal singular subspaces and half-spin Grassmannians (Section 4.8). In Section 4.9 we describe all apartments preserving mappings of polar and half-spin Grassmannians; in particular, it will be shown that all apartments preserving bijections are collineations of the corresponding Grassmann spaces.

4.1 Polar spaces

4.1.1 *Axioms and elementary properties*

Following [Buekenhout and Shult (1974)] we define a *polar space* (of finite rank) as a partial linear space $\Pi = (P, \mathcal{L})$ satisfying the following axioms:

(1) each line contains at least 3 points,
(2) if $p \in P$ and $L \in \mathcal{L}$ then p is collinear with one or all points of the line L (the Buekenhout–Shult property),
(3) there is no point collinear with all other points,
(4) every flag consisting of singular subspaces is finite (this implies that every singular subspace is finite-dimensional).

The collinearity relation will be denoted by $\perp$: we write $p \perp q$ if p and q are collinear points, and $p \not\perp q$ otherwise. More general, $X \perp Y$ means that every point of X is collinear with every point of Y. For every subset $X \subset P$ we define

$$X^\perp := \{\, p \in P \, : \, p \perp X \,\}.$$

The axiom (2) guarantees that $X^\perp$ is a subspace of Π; moreover, for every point $p \in P$ the subspace $p^\perp$ is a hyperplane of Π (recall that a proper subspace of a partial linear space is called a hyperplane if it has a non-empty intersection with every line).

It follows from the axiom (2) that polar spaces are connected gamma spaces and the distance between non-collinear points is equal to 2.

Let $X \subset P$ be a subset satisfying $X \perp X$ (a clique of the collinearity graph of Π). By Section 2.5, X is contained in a certain singular subspace of Π. Recall that the minimal singular subspace containing X is called *spanned* by X and denoted by $\langle X \rangle$. Since for every point $p \in P$

$$p \perp X \implies p \perp \langle X \rangle$$

(Corollary 2.3), we have $X^\perp = \langle X \rangle^\perp$.

If S is a singular subspace of Π and $p \not\perp S$ then the axiom (2) implies that $S \cap p^\perp$ is a hyperplane of S (hyperplanes of a line are points). If S is a maximal singular subspace then $S^\perp = S$ (Π is a gamma space and, by Proposition 2.7, S is a maximal clique of the collinearity graph); in this case, $S \cap p^\perp$ is a hyperplane of S for every point $p \in P \setminus S$.

Theorem 4.1 ([Buekenhout and Shult (1974)]). *Suppose that a polar space $\Pi = (P, \mathcal{L})$ contains a singular subspace of dimension greater than 1. Then the following assertions are fulfilled:*

(1) *all maximal singular subspaces are projective spaces of the same finite dimension,*

(2) *for every maximal singular subspace S there exists a maximal singular subspace disjoint from S.*

Remark 4.1. Polar spaces of rank $n \geq 3$ were defined in [Tits (1974)] and [Veldkamp (1959/1960)] as partial linear spaces satisfying the following axioms:

- all maximal singular subspaces are $(n-1)$-dimensional projective spaces,
- for every maximal singular subspace S and every point $p \notin S$, all points of S collinear with p form a hyperplane of S,
- there exist two disjoint maximal singular subspaces.

It follows from Theorem 4.1 that the Tits–Veldkamp and Buekenhout–Shult definitions of polar spaces are equivalent.

The original proof given by F. Buekenhout and E. Shult was rather complicated. In [Buekenhout (1990)] Theorem 4.1 was drawn from the following result.

Theorem 4.2 ([Teirlinck (1980)]). *Suppose that a linear space has a family of hyperplanes $\mathcal{H}$ which satisfies the following conditions:*

(a) *for every distinct hyperplanes $H_1, H_2 \in \mathcal{H}$ and every point p there is a hyperplane $H \in \mathcal{H}$ containing $H_1 \cap H_2$ and p,*

(b) *for every point p there exists a hyperplane $H \in \mathcal{H}$ which does not contain p.*

If every line contains at least 3 points then the linear space is a projective space.

4.1.2 *Proof of Theorem 4.1*

Lemma 4.1. *Every hyperplane in a linear space is a maximal proper subspace of this linear space.*

Proof. If H is a hyperplane of a linear space then this linear space is spanned by H and any point $p \notin H$ (since for every point $q \neq p$ the line pq intersects H). □

Proposition 4.1. *Every singular subspace of* Π *whose dimension is not less than 2 is a projective space.*

Proof. Clearly, we can restrict ourselves to maximal singular subspaces. Let S be a maximal singular subspace whose dimension is assumed to be not less than 2. It was noted above that

$$H_p := S \cap p^{\perp}$$

is a hyperplane of S for every point $p \in P \setminus S$. Denote by S_p the singular subspace spanned by H_p and p. If $q \in S_p \setminus H_p$ then H_q coincides with H_p (it is clear that $H_p \subset H_q$ and Lemma 4.1 gives the claim). We show that the family of hyperplanes

$$\{H_p\}_{p \in P \setminus S}$$

satisfies the conditions of Theorem 4.2.

(a). Let H_p and H_q be distinct hyperplanes of S. Then $p \notin S_q$ and $q \notin S_p$. We take any point $t \in H_q \setminus H_p$. Since $t \not\perp p$, the line tq contains a unique point q' collinear with p. This point does not belong to H_q. Therefore, $q' \in S_q \setminus H_q$. Suppose that the line pq' intersects S in a certain point. This point belongs to the intersection of H_p and $H_{q'} = H_q$. This means that the line pq' is contained in S_q (since S_q contains q' and $H_p \cap H_q$) which contradicts the fact that $p \notin S_q$. Thus the line pq' does not intersect S. Each point of pq' is collinear with all points of $H_p \cap H_q$. For every point $u \in S$ we can choose a point $v \in pq'$ collinear with u. The hyperplane H_v contains u and $H_p \cap H_q$.

(b). For every point $u \in S$ the axiom (3) implies the existence of a point $p \in P \setminus S$ which is not collinear with u. The associated hyperplane H_p does not contain u. $\square$

Proposition 4.2. *All maximal singular subspaces of* Π *have the same dimension.*

Proof. Let S and U be maximal singular subspaces of dimension n and k (respectively) and $k \le n$. Suppose that the dimension of complements of $S \cap U$ in U is equal to m (see Remark 4.2). Then

$$\dim(S \cap U) = k - m - 1.$$

Let $t_1, \ldots, t_{m+1}$ be a base of a certain complement of $S \cap U$ in U. The subspace $S \cap U^{\perp}$ is the intersection of the hyperplanes

$$S \cap t_1^{\perp}, \ldots, S \cap t_{m+1}^{\perp}$$

and

$$\dim(S \cap U^{\perp}) \geq n - m - 1.$$

Since $U = U^{\perp}$ (U is a maximal singular subspace), we get

$$k - m - 1 \geq n - m - 1$$

and $k \geq n$ which implies $k = n$. $\square$

Remark 4.2. Let M, N, T be subspaces of a projective space such that $M, N \subset T$. We say that N is a *complement* of M in T if

$$M \cap N = \emptyset \ \text{ and } \ \langle M, N \rangle = T.$$

Then

$$\dim M + \dim N + 1 = \dim T$$

(for projective planes this is trivial; for projective spaces whose dimension is greater than 2 this follows from Theorem 1.3).

So, if a polar space contains a singular subspace of dimension greater than 1 then all maximal singular subspaces are projective spaces of a certain finite dimension $n \geq 2$; the number $n + 1$ is said to be the *rank* of this polar space. If a polar space does not satisfy the condition of Theorem 4.1 then all maximal singular subspaces are lines, and we say that it is a *polar space of rank* 2 or a *generalized quadrangle*.

Lemma 4.2. *Let S be a maximal singular subspace in a polar space of rank n. Let also U be a singular subspace such that the dimension of complements of $S \cap U$ in U is equal to m. Then*

$$\dim(S \cap U^{\perp}) = n - m - 2$$

and

$$\langle U, S \cap U^{\perp} \rangle \tag{4.1}$$

is a maximal singular subspace containing U.

Proof.   As in the proof of Proposition 4.2, we establish that

$$\dim(S \cap U^{\perp}) \geq n - m - 2.$$

The subspace (4.1) is spanned by $S \cap U^{\perp}$ and a compliment of $S \cap U$ in U. Since these subspaces are disjoint and the latter subspace is m-dimensional, the dimension of (4.1) is not less than $n - 1$. This means that (4.1) is a maximal singular subspace and its dimension is equal to $n - 1$. The latter guarantees that $S \cap U^{\perp}$ is $(n - m - 2)$-dimensional. $\square$

Proposition 4.3. *For every maximal singular subspace S there exists a maximal singular subspace disjoint from S.*

Proof. Let S and U be maximal singular subspaces such that $S \cap U \neq \emptyset$. By the axiom (3), there exists a point p non-collinear with a certain point $t \in S \cap U$. Clearly, $p \notin U$ and we denote by U' the maximal singular subspace spanned by p and $U \cap p^\perp$ (Lemma 4.2). If U' contains a point $q \in S \setminus U$ then

$$U \cap p^\perp = U \cap q^\perp$$

which is impossible ($t \in S \cap U$ is collinear with q and non-collinear with p). Therefore, $S \cap U'$ is contained in $S \cap U$. Since $t \notin U'$,

$$\dim(S \cap U') < \dim(S \cap U).$$

Step by step, we construct a maximal singular subspace disjoint from S. $\square$

Theorem 4.1 is the union of Propositions 4.2 and 4.3.

4.1.3 Corollaries of Theorem 4.1

Let $\Pi = (P, \mathcal{L})$ be a polar space of rank n.

Proposition 4.4. *Every non-maximal singular subspace of Π can be presented as the intersection of two maximal singular subspaces.*

Proof. Let S be a singular subspace and U be a maximal singular subspace containing S. By Theorem 4.1, there exists a maximal singular subspace U' disjoint from U. Consider the maximal singular subspace

$$U'' := \langle S, U' \cap S^\perp \rangle$$

(Lemma 4.2). Note that S and $U' \cap S^\perp$ are disjoint. The subspace $U \cap U''$ does not intersect $U' \cap S^\perp$ and the inclusion $S \subset U \cap U''$ guarantees that $U \cap U''$ coincides with S. $\square$

If S is a singular subspace of Π then $S^\perp$ is the union of all maximal singular subspaces containing S (recall that Π is a gamma space and the class of maximal singular subspaces coincides with the class of maximal cliques of the collinearity graph). By Proposition 4.4, for any pair of singular subspaces S and U the inclusion $S^\perp \subset U^\perp$ implies that $U \subset S$; hence, $S^\perp = U^\perp$ if and only if $S = U$.

Now suppose that X is a clique of the collinearity graph of Π. As above, $X^\perp$ is the union of all maximal singular subspaces containing X and $X^{\perp\perp}$

is the intersection of all these subspaces. Then $\langle X \rangle \subset X^{\perp\perp}$ and Proposition 4.4 implies that

$$X^{\perp\perp} = \langle X \rangle.$$

As a consequence, we obtain the following characterization of lines in terms of the collinearity relation: if p and q are distinct collinear points then

$$p\,q = \{p, q\}^{\perp\perp}.$$

This means that every isomorphism between the collinearity graphs of polar spaces is a collineation between these polar spaces.

We will need the following result concerning pairs of non-collinear points in polar spaces.

Lemma 4.3. *If $n \geq 3$ and p, q are non-collinear points of Π then the subspace $p^{\perp} \cap q^{\perp}$ is a polar space of rank $n - 1$.*

Proof. The axioms (1), (2), (4) are trivial. We verify (3).

Suppose that there exists a point $t \in p^{\perp} \cap q^{\perp}$ collinear with all points of $p^{\perp} \cap q^{\perp}$. For every point $s \in p^{\perp} \setminus \{p\}$ the line $p\,s$ contains two distinct points collinear with t (one of these point is p and the other is the intersection with the hyperplane $q^{\perp}$); by the axiom (2), t is collinear with s. Thus $p^{\perp} \subset t^{\perp}$ and $p = t$ which contradicts $p \notin p^{\perp} \cap q^{\perp}$.

So, $p^{\perp} \cap q^{\perp}$ is a polar space. It is clear that $p^{\perp} \cap q^{\perp}$ does not contain $(n - 1)$-dimensional singular subspaces. Every maximal singular subspace containing p intersects $q^{\perp}$ in an $(n - 2)$-dimensional subspace. Therefore, our polar space is of rank $n - 1$. $\square$

4.1.4 *Polar frames*

Let $\Pi = (P, \mathcal{L})$ be a polar space of rank n. We say that a subset

$$\{p_1, \ldots, p_{2n}\}$$

is a *frame* of Π if for every $i \in \{1, \ldots, 2n\}$ there is unique $\sigma(i) \in \{1, \ldots, 2n\}$ such that

$$p_i \not\perp p_{\sigma(i)}.$$

First of all we show that frames exist.

Let S and U be disjoint maximal singular subspaces of Π. We take any base $B = \{p_1, \ldots, p_n\}$ of S. By Lemma 4.2, for every $i \in \{1, \ldots, n\}$ there is a unique point of U collinear with all points of $B \setminus \{p_i\}$; we denote it by p_{n+i}. This point is non-collinear with p_i (otherwise $p_{n+i} \perp S$ and the maximal singular subspace S contains p_{n+i} which is impossible, since $S \cap U = \emptyset$). Then $\{p_1, \ldots, p_{2n}\}$ is a frame of Π.

Proposition 4.5. *Every frame is an independent subset.*

Proof. If $B = \{p_1, \ldots, p_{2n}\}$ is a frame of Π then $B \setminus \{p_i\}$ is contained in the hyperplane $p_{\sigma(i)}^{\perp}$; but this hyperplane does not contain p_i. $\square$

Remark 4.3. In some cases, frames are not bases of a polar space (examples will be given in Subsection 4.3.1).

Since every subset of an independent subset is independent, any k distinct mutually collinear points in a frame span a $(k-1)$-dimensional singular subspace.

Proposition 4.6. *Let B be a frame and S, U be singular subspaces spanned by subsets of B. Then $S \cap U$ is spanned by the set $S \cap U \cap B$. In particular, if this set is empty then S and U are disjoint.*

Proof. Suppose that $B = \{p_1, \ldots, p_{2n}\}$. Since B is an independent subset, the subspaces S and U are spanned by $S \cap B$ and $U \cap B$, respectively. First, we establish that

$$(S \cap B) \cap (U \cap B) = \emptyset \implies S \cap U = \emptyset.$$

An easy verification shows that there exist disjoint subsets $X, Y \subset B$ such that $\langle X \rangle, \langle Y \rangle$ are maximal singular subspaces and

$$S \cap B \subset X, \quad U \cap B \subset Y.$$

Suppose that $\langle X \rangle$ and $\langle Y \rangle$ have a non-empty intersection and consider a point p belonging to $\langle X \rangle \cap \langle Y \rangle$. The intersection of all

$$\langle X \setminus \{p_i\} \rangle, \quad p_i \in X,$$

is empty and there exists $p_i \in X$ such that $\langle X \setminus \{p_i\} \rangle$ does not contain p. Then

$$(X \setminus \{p_i\}) \cup \{p\}$$

is a base of $\langle X \rangle$. The point $p_{\sigma(i)} \in B \setminus X = Y$ is collinear with all points of this base (it is trivial that $p_{\sigma(i)} \perp X \setminus \{p_i\}$ and we have $p_{\sigma(i)} \perp p$, since p and $p_{\sigma(i)}$ belong to $\langle Y \rangle$). This means that $p_{\sigma(i)} \perp \langle X \rangle$ which contradicts the fact that $p_i \in \langle X \rangle$. Therefore, $\langle X \rangle$ and $\langle Y \rangle$ are disjoint. Since $S \subset \langle X \rangle$ and $U \subset \langle Y \rangle$, we get the claim.

Now suppose that $S \cap U \cap B \neq \emptyset$. Then S is spanned by the subspaces

$$\langle S \cap U \cap B \rangle \text{ and } \langle (B \cap S) \setminus (U \cap B) \rangle.$$

By the first part of our proof, these subspaces are disjoint and the second subspace is disjoint from U; hence it does not intersects $S \cap U$. The inclusion

$$\langle S \cap U \cap B \rangle \subset S \cap U \subset S$$

guarantees that $\langle S \cap U \cap B \rangle$ coincides with $S \cap U$. □

Corollary 4.1. *If B is a frame of Π then there is no point of Π collinear with all points of B.*

Proof. A point collinear with all points of B is contained in every maximal singular subspace spanned by a subset of B. By Proposition 4.6, there exist disjoint maximal singular subspaces spanned by subsets of B. □

Proposition 4.7. *For any singular subspaces S and U there is a frame of Π such that S and U are spanned by subsets of this frame.*

Proof. We prove the statement induction by n. The case $n = 2$ is trivial and we suppose that $n \geq 3$.

If $S \perp U$ then there exists a maximal singular subspace M containing S and U. We choose a base of M such that S and U are spanned by subsets of this base. It was shown above that this base can be extended to a frame of Π.

Now suppose that $S \not\perp U$. In this case, there are non-collinear points $p \in S$ and $q \in U$. The singular subspaces

$$S' := S \cap q^{\perp} \text{ and } U' := U \cap p^{\perp}$$

are contained in the polar space $p^{\perp} \cap q^{\perp}$ (Lemma 4.3). By the inductive hypothesis, there exists a frame B' of $p^{\perp} \cap q^{\perp}$ such that S' and U' are spanned by subsets of B'. Since

$$S = \langle S', p \rangle \text{ and } U = \langle U', q \rangle,$$

the frame $B' \cup \{p, q\}$ is as required. □

Corollary 4.2. *For any singular subspaces S_1, S_2 there exist maximal singular subspaces M_1, M_2 such that*

$$M_1 \cap M_2 = S_1 \cap S_2$$

and $S_i \subset M_i$ for $i = 1, 2$.

Proof. Consider a frame B such that S_1 and S_2 are spanned by subsets of B. There exists a maximal singular subspace

$$M_1 = \langle X \rangle, \quad X \subset B,$$

containing S_1 and intersecting S_2 precisely in $S_1 \cap S_2$. If $M = \langle B \setminus X \rangle$ then the maximal singular subspace

$$M_2 := \langle S_2, M \cap S_2^\perp \rangle$$

contains S_2 and intersects M_1 precisely in $S_1 \cap S_2$. $\qquad\square$

4.2 Grassmannians

4.2.1 *Polar Grassmannians*

As above, we suppose that $\Pi = (P, \mathcal{L})$ is a polar space of rank n. For every $k \in \{0, 1, \dots, n-1\}$ denote by $\mathcal{G}_k(\Pi)$ the *polar Grassmannian* consisting of all k-dimensional singular subspaces of Π. Then $\mathcal{G}_0(\Pi) = P$.

In the case when $k \le n-2$, we say that $S, U \in \mathcal{G}_k(\Pi)$ are *adjacent* if $S \perp U$ and their intersection is $(k-1)$-dimensional (or, equivalently, if S and U span a $(k+1)$-dimensional singular subspace). If S and U are distinct elements of $\mathcal{G}_{n-1}(\Pi)$ then $S \not\perp U$; such subspaces are said to be *adjacent* if their intersection belongs to $\mathcal{G}_{n-2}(\Pi)$.

Let M and N be incident singular subspaces of Π such that

$$\dim M < k < \dim N.$$

As in Section 3.1, we define

$$[M, N]_k := \{\, S \in \mathcal{G}_k(\Pi) \ : \ M \subset S \subset N \,\};$$

if $M = \emptyset$ then we will write $\langle N]_k$ instead of $[M, N]_k$. Also denote by $[M\rangle_k$ the set of all elements of $\mathcal{G}_k(\Pi)$ containing M. In the case when $0 \le k < n-1$, we say that $[M, N]_k$ is a *line* of $\mathcal{G}_k(\Pi)$ if

$$\dim M = k-1 \quad \text{and} \quad \dim N = k+1.$$

The set $[M\rangle_{n-1}$ is said to be a *line* of $\mathcal{G}_{n-1}(\Pi)$ if M belongs to $\mathcal{G}_{n-2}(\Pi)$. The set of all lines of $\mathcal{G}_k(\Pi)$ will be denoted by $\mathcal{L}_k(\Pi)$.

Two distinct elements of $\mathcal{G}_k(\Pi)$ are joined by a line if and only if they are adjacent. For any adjacent $S, U \in \mathcal{G}_k(\Pi)$ there is precisely one line containing them:

$$[S \cap U, \langle S, U \rangle]_k \quad \text{if } k \le n-2$$

and $[S \cap U\rangle_k$ if $k = n-1$.

Exercise 4.1. Show that every line of $\mathcal{G}_k(\Pi)$, $k < n-1$, contains at least three distinct points.

The pairs

$$\mathfrak{G}_k(\Pi) := (\mathcal{G}_k(\Pi), \mathcal{L}_k(\Pi)), \quad k \in \{0, 1, \ldots, n-1\},$$

are partial linear spaces; they are called the *Grassmann spaces* of Π. It is clear that $\mathfrak{G}_0(\Pi) = \Pi$. The Grassmann space $\mathfrak{G}_{n-1}(\Pi)$ also is known as the *dual polar space* of Π [Cameron (1982)].

Proposition 4.8. *The Grassmann space $\mathfrak{G}_k(\Pi)$ is connected for every k. The distance between $S, U \in \mathcal{G}_{n-1}(\Pi)$ is equal to*

$$n - 1 - \dim(S \cap U).$$

Remark 4.4. In the general case, the distance formula is more complicated.

Proof. Let $S, U \in \mathcal{G}_k(\Pi)$. We define

$$\mathrm{cd}(S, U) := k - \dim(S \cap U)$$

(if $k = n - 1$ then $\mathrm{cd}(S, U) = 1$ is equivalent to the fact that S and U are adjacent).

Suppose that $k = n - 1$ and $\mathrm{cd}(S, U) > 1$. We take any point $p \in U \setminus S$ and denote by S_1 the maximal singular subspace spanned by $S \cap p^\perp$ and p. Then S_1 is adjacent with S and

$$\mathrm{cd}(S_1, U) = \mathrm{cd}(S, U) - 1.$$

Step by step, we construct a sequence of maximal singular subspaces

$$S = S_0, S_1, \ldots, S_i = U, \quad i = \mathrm{cd}(S, U),$$

such that S_{j-1} and S_j are adjacent for every $j \in \{1, \ldots, i\}$. As in the proof of Proposition 3.1, for every path

$$S = U_0, U_1, \ldots, U_l = U$$

in the collinearity graph of $\mathfrak{G}_{n-1}(\Pi)$ we have

$$\dim(U_0 \cap U_1 \cap \cdots \cap U_l) \geq (n-1) - l.$$

The trivial inclusion

$$U_0 \cap U_1 \cap \cdots \cap U_l \subset S \cap U$$

guarantees that

$$\dim(S \cap U) \geq (n-1) - l.$$

Thus $l \geq \mathrm{cd}(S, U)$ which means that the distance between S and U is equal to $\mathrm{cd}(S, U)$.

Let $k \leq n-2$ and S, U be non-adjacent elements of $\mathcal{G}_k(\Pi)$. If $S \perp U$ then S, U are contained in the singular subspace $\langle S \cup U \rangle$ and a path connecting S with U can be constructed as in the proof of Proposition 3.1. In the case when $S \not\perp U$, we prove the connectedness induction by $\mathrm{cd}(S, U)$.

If $\mathrm{cd}(S, U) = 1$ then there exists a frame $\{p_1, \ldots, p_{2n}\}$ of Π such that
$$S = \langle p_{i_1}, \ldots, p_{i_k}, p_m \rangle \quad \text{and} \quad U = \langle p_{i_1}, \ldots, p_{i_k}, p_{\sigma(m)} \rangle.$$
We choose
$$j \in \{1, \ldots, 2n\} \setminus \{i_1, \sigma(i_1), \ldots, i_k, \sigma(i_k), m, \sigma(m)\}$$
(this is possible, since $k \leq n - 2$). The k-dimensional singular subspace
$$\langle p_{i_1}, \ldots, p_{i_k}, p_j \rangle$$
is adjacent with both S and U.

In the case when $\mathrm{cd}(S, U) > 1$, we take a point $p \in U$ satisfying $p \not\perp S$ and write S_2 for the k-dimensional singular subspace spanned by $S \cap p^\perp$ and p. Then
$$\mathrm{cd}(S, S_2) = 1 \quad \text{and} \quad \mathrm{cd}(S_2, U) = \mathrm{cd}(S, U) - 1.$$
There exist $S_1 \in \mathcal{G}_k(\Pi)$ adjacent with both S, S_2 and the inductive hypothesis implies the existence of a path connecting S_2 with U. $\qquad\square$

Lemma 4.4. *Let $0 < k < n - 1$ and U be a $(k-1)$-dimensional singular subspace of Π. The following assertions are fulfilled:*

(1) *$[U\rangle_k$ is a polar space of rank $n - k$,*

(2) *for every frame $\mathcal{B}$ of $[U\rangle_k$ there is a frame B of Π such that U is spanned by a subset of B and*
$$\mathcal{B} = \mathcal{A} \cap [U\rangle_k,$$
where $\mathcal{A}$ consists of all k-dimensional singular subspaces spanned by subsets of B.

Proof. (1). An easy verification shows that $[U\rangle_k$ is a polar space and every its maximal singular subspace is $[U, M]_k$, where M is a maximal singular subspace of Π. The latter means that the rank of our polar space is equal to $n - k$.

(2). Let $\{S_1, \ldots, S_{2n-2k}\}$ be a frame of $[U\rangle_k$. In each S_i we take a point $p_i \in S_i \setminus U$ and denote by X the set of all p_i. For every p_i there is unique $p_{\sigma(i)}$ such that $p_i \not\perp p_{\sigma(i)}$. By Lemma 4.3, $X^\perp$ is a polar space of rank k. This polar space contains U and we choose a frame Y of $X^\perp$ such that U is spanned by a subset of Y. Then $X \cup Y$ is a frame of Π satisfying the required conditions. $\qquad\square$

4.2.2   *Two types of polar spaces*

Theorem 4.3. *For a rank n polar space one of the following possibilities is realized:*

(C) *every $(n-2)$-dimensional singular subspace is contained in at least three distinct maximal singular subspaces,*

(D) *every $(n-2)$-dimensional singular subspace is contained in precisely two maximal singular subspaces.*

We say that a rank n polar space is of type C_n or D_n if the corresponding case is realized.

Proof.   Let $\Pi = (P, \mathcal{L})$ be a polar space of rank n. First consider the case when $n = 2$.

Suppose that $\{p_1, p_2, p_3, p_4\}$ is a frame of Π where

$$p_1 \not\perp p_3 \ \text{ and } \ p_2 \not\perp p_4.$$

Every point p_i lies on precisely two lines from the collection

$$p_1 p_2, \ p_1 p_4, \ p_2 p_3, \ p_3 p_4.$$

Let L be a third line passing through p_1.

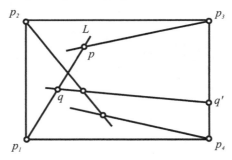

There is a point $p \in L$ collinear with p_3 and pp_3 is a third line passing through p_3. Now consider a point $q \in L \setminus \{p_1, p\}$ and take a unique point $q' \in p_3 p_4$ collinear with q (it is clear that the point q' is distinct from p_3 and p_4). A point on the line qq' collinear with p_2 gives a third line passing through p_2. This line contains a point collinear with p_4 and we get a third line through p_4. Therefore, for every $i \in \{1, 2, 3, 4\}$ there are at least three distinct lines passing through p_i. Since any pair of points is contained in a certain frame, the polar space Π is of type C_2 if there is a point belonging to at least three distinct lines.

Now suppose that $n \geq 3$. We need to prove the following: if a certain $S \in \mathcal{G}_{n-2}(\Pi)$ is contained in at least three distinct maximal singular subspaces then the same holds for all elements of $\mathcal{G}_{n-2}(\Pi)$.

Let N be an $(n-3)$-dimensional subspace of S. By Lemma 4.4, $[N\rangle_{n-2}$ is a generalized quadrangle. Since S (as a point of this generalized quadrangle) lies on at least three distinct lines, the generalized quadrangle is of type C_2. This means that every element of $[N\rangle_{n-2}$ is contained in at least three distinct maximal singular subspaces.

Therefore, $U \in \mathcal{G}_{n-2}(\Pi)$ is contained in at least three distinct maximal singular subspaces if it has an $(n-3)$-dimensional intersection with S, in particular, if S and U are adjacent. The connectedness of the Grassmann space $\mathfrak{G}_{n-2}(\Pi)$ gives the claim. □

Every line of $\mathfrak{G}_{n-1}(\Pi)$ contains at least three distinct points if Π is a polar space of type C_n. In the case when Π is of type D_n, the Grassmann space $\mathfrak{G}_{n-1}(\Pi)$ is trivial: every line consists of two points; in other words, lines are edges of the collinearity graph.

Remark 4.5. If Π is a generalized quadrangle of type C_2 then $\mathfrak{G}_1(\Pi)$ is a generalized quadrangle of type C_2. If Π is a generalized quadrangle of type D_2 then $\mathfrak{G}_1(\Pi)$ does not satisfy the polar axiom (1).

Exercise 4.2. Let Π be a generalized quadrangle of type D_2 and $\{p_1, p_2, q_1, q_2\}$ be a frame of Π such that $p_i \not\perp q_i$ for $i \in \{1, 2\}$. Show that every line of Π intersects the lines $p_1 p_2$ and $q_1 q_2$ or the lines $p_1 q_2$ and $p_2 q_1$. Therefore, if every line of Π consists of three points then Π looks as in the picture below.

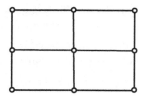

Proposition 4.9. *Every frame in a polar space of type D_n is a base of this polar space.*

Proof. Let $\Pi = (P, \mathcal{L})$ be a polar space of type D_n and

$$B = \{p_1, \ldots, p_n, q_1, \ldots, q_n\}$$

be a frame of Π such that $p_i \not\perp q_i$ for every $i \in \{1, \ldots, n\}$. By Proposition 4.5, B is an independent subset and we need to show that Π is spanned by

B. In the case when $n = 2$, this follows from Exercise 4.2. Suppose that $n \geq 3$ and prove the statement induction by n.

By the inductive hypothesis, the polar space $p_1^\perp \cap q_1^\perp$ is spanned by $B \setminus \{p_1, q_1\}$. For every point $t \in p_1^\perp \setminus \{p_1\}$ the line $p_1 t$ intersects $p_1^\perp \cap q_1^\perp$ and $p_1^\perp \subset \langle B \rangle$. The same inclusion holds for all points of B. Suppose that $p \in P$ is non-collinear with every point of B. We take $x \in p_1 p_2$ collinear with p and $y \in q_1 q_2$ collinear with x. Then

$$B' = (B \setminus \{p_1, q_2\}) \cup \{x, y\}$$

is a frame of Π and $p \in \langle B' \rangle = \langle B \rangle$. □

4.2.3 *Half-spin Grassmannians*

Throughout this subsection we suppose that $\Pi = (P, \mathcal{L})$ is a polar space of type D_n. We show that the Grassmannian $\mathcal{G}_{n-1}(\Pi)$ can be uniquely decomposed in two disjoint parts such that the distance between two elements of $\mathcal{G}_{n-1}(\Pi)$ is odd if and only if these elements belong to the different parts. Recall that

$$d(S, U) = n - 1 - \dim(S \cap U).$$

for all $S, U \in \mathcal{G}_{n-1}(\Pi)$.

Lemma 4.5. *Let S and U be adjacent elements of $\mathcal{G}_{n-1}(\Pi)$. Then for any $N \in \mathcal{G}_{n-1}(\Pi)$ the distance $d(S, N)$ is odd if and only if $d(U, N)$ is even.*

Proof.   The statement is trivial if N coincides with S or U and we assume that N is distinct from S and U. Observe that every point of $S \setminus U$ is non-collinear with every point of $U \setminus S$. This means that at least one of the subspaces $S \cap N, U \cap N$ is contained in $S \cap U$ (otherwise N contains points $x \in S \setminus U$ and $y \in U \setminus S$ which cannot be collinear).

Lemma 4.2 implies the existence of a point

$$p \in N \setminus (S \cap U)$$

collinear with all points of $S \cap U$. If $S \cap N$ and $U \cap N$ both are contained in $S \cap U$ then p does not belong to $S \cup U$ and $\langle S \cap U, p \rangle$ is a third maximal singular subspace containing $S \cap U$ which contradicts the assumption that Π is of type D_n.

Therefore, only one of these subspaces is contained in $S \cap U$. Consider the case when

$$S \cap N \subset S \cap U \quad \text{and} \quad U \cap N \not\subset S \cap U.$$

Then

$$S \cap N = S \cap U \cap N$$

is a hyperplane of $U \cap N$ (since $S \cap U$ is a hyperplane of U). This implies that

$$d(S, N) = d(U, N) + 1.$$

The second case is similar. □

Theorem 4.4. *There is a unique pair of disjoint subsets*

$$\mathcal{X}, \mathcal{Y} \subset \mathcal{G}_{n-1}(\Pi)$$

satisfying the following conditions:

$$\mathcal{X} \cup \mathcal{Y} = \mathcal{G}_{n-1}(\Pi),$$

the distance between two elements of $\mathcal{G}_{n-1}(\Pi)$ is odd if and only if one of them belongs to $\mathcal{X}$ and other belongs to $\mathcal{Y}$.

Proof. We fix $N \in \mathcal{G}_{n-1}(\Pi)$ and define

$$\mathcal{X} := \{S \in \mathcal{G}_{n-1}(\Pi) \ : \ d(S, N) \text{ is even}\},$$

$$\mathcal{Y} := \{U \in \mathcal{G}_{n-1}(\Pi) \ : \ d(U, N) \text{ is odd}\}.$$

These are disjoint subsets whose union is $\mathcal{G}_{n-1}(\Pi)$. By the previous lemma, any two adjacent elements of $\mathcal{G}_{n-1}(\Pi)$ belong to the different subsets. Therefore, if the distance between two elements of $\mathcal{G}_{n-1}(\Pi)$ is odd then one of them belongs to $\mathcal{X}$ and the other belongs to $\mathcal{Y}$; in the case when the distance is even, the elements both belong to $\mathcal{X}$ or $\mathcal{Y}$.

Suppose that

$$\mathcal{X}', \mathcal{Y}' \subset \mathcal{G}_{n-1}(\Pi)$$

is another pair of subsets satisfying the same conditions. Then $\mathcal{X}'$ intersects at least one of the subsets $\mathcal{X}, \mathcal{Y}$. Let $\mathcal{X} \cap \mathcal{X}' \neq \emptyset$ and S be an element of this intersection. Then $d(S, U)$ is even for every U belonging to $\mathcal{X} \cup \mathcal{X}'$. This means that $\mathcal{X}' = \mathcal{X}$, hence $\mathcal{Y}' = \mathcal{Y}$. □

The subsets described in Theorem 4.4 will be denoted by

$$\mathcal{G}_+(\Pi), \ \ \mathcal{G}_-(\Pi)$$

and called the *half-spin Grassmannians* of Π. The distance between any two elements of the half-spin Grassmannian (in the collinearity graph of $\mathfrak{G}_{n-1}(\Pi)$) is even.

Let $\delta \in \{+, -\}$. Two elements of $\mathcal{G}_\delta(\Pi)$ are said to be *adjacent* if their intersection is $(n-3)$-dimensional (the distance in the collinearity graph of $\mathfrak{G}_{n-1}(\Pi)$ is equal to 2). The intersection

$$[M\rangle_\delta := [M\rangle_{n-1} \cap \mathcal{G}_\delta(\Pi)$$

is called a *line* of $\mathcal{G}_\delta(\Pi)$ if M is $(n-3)$-dimensional. The set of all such lines will be denoted by $\mathcal{L}_\delta(\Pi)$.

If $n = 2$ then any two distinct elements of $\mathcal{G}_\delta(\Pi)$ are disjoint and there is only one line which coincides with $\mathcal{G}_\delta(\Pi)$. In what follows we will restrict ourselves to the case when $n \geq 3$.

Two distinct elements of $\mathcal{G}_\delta(\Pi)$ are joined by a line if and only if they are adjacent. If $S, U \in \mathcal{G}_\delta(\Pi)$ are adjacent then $[S \cap U\rangle_\delta$ is the unique line containing them.

If $\delta \in \{+, -\}$ then we write $-\delta$ for the sing satisfying $\{\delta, -\delta\} = \{+, -\}$.

Proposition 4.10. *Every line of $\mathcal{G}_\delta(\Pi)$, $\delta \in \{+, -\}$, contains at least three points.*

Proof. Let $M \in \mathcal{G}_{n-3}(\Pi)$. Consider any $(n-2)$-dimensional singular subspace N_1 containing M. By Proposition 4.4, this is the intersection of two maximal singular subspaces S_1 and U_1; one of them belongs to $\mathcal{G}_\delta(\Pi)$ and the other is an element of $\mathcal{G}_{-\delta}(\Pi)$. Suppose that $U_1 \in \mathcal{G}_{-\delta}(\Pi)$. We choose two distinct $(n-2)$-dimensional singular subspaces $N_2, N_3 \neq N_1$ contained in U_1 and containing M (in other words, N_1, N_2, N_3 are distinct points on the line $[M, U_1]_{n-2}$). Proposition 4.4 implies the existence of $S_i \in \mathcal{G}_\delta(\Pi)$ ($i = 2, 3$) intersecting U_1 precisely in N_i. Then S_1, S_2, S_3 are distinct points on the line $[M\rangle_\delta$. $\square$

The patrial linear spaces

$$\mathfrak{G}_\delta(\Pi) := (\mathcal{G}_\delta(\Pi), \mathcal{L}_\delta(\Pi)), \quad \delta \in \{+, -\},$$

are called the *half-spin Grassmann spaces* of Π.

Exercise 4.3. Shows that every path of length j in the collinearity graph of $\mathfrak{G}_\delta(\Pi)$, $\delta \in \{+, -\}$, can be extended to a path of length $2j$ in the collinearity graph of $\mathfrak{G}_{n-1}(\Pi)$. *Hint:* if $S, U \in \mathcal{G}_\delta(\Pi)$ are adjacent then there exist $p \in S \setminus U$ and $q \in U \setminus S$ satisfying $p \perp q$.

Proposition 4.11. *The half-spin Grassmann space $\mathfrak{G}_\delta(\Pi)$, $\delta \in \{+, -\}$, is connected. The distance between $S, U \in \mathcal{G}_\delta(\Pi)$ (in the collinearity graph of $\mathfrak{G}_\delta(\Pi)$) is equal to*

$$\frac{n - 1 - \dim(S \cap U)}{2}.$$

Proof. Let $S, U \in \mathcal{G}_\delta(\Pi)$. By Proposition 4.8, the collinearity graph of $\mathfrak{G}_{n-1}(\Pi)$ contains a path

$$S = S_0, S_1, \ldots, S_i = U, \quad i = n - 1 - \dim(S \cap U).$$

Theorem 4.4 guarantees that i is even and

$$S_0, S_2, \ldots, S_{i-2}, S_i$$

is a path in the collinearity graph of $\mathfrak{G}_\delta(\Pi)$. The distance formula follows immediately from Exercise 4.3. $\qquad\square$

4.3 Examples

4.3.1 *Polar spaces associated with sesquilinear forms*

Let V be an n-dimensional left vector space over a division ring R and Ω be a non-degenerate reflexive form defined on V. Suppose that Ω has totally isotropic subspaces of dimensions 2. This implies that $n \geq 4$ (the dimension of a totally isotropic subspace is not greater than the codimension). We write $\mathcal{L}(\Omega)$ for the set formed by all lines of Π_V such that the associated 2-dimensional linear subspaces are totally isotropic. Distinct $P, P' \in \mathcal{G}_1(\Omega)$ are joined by a such line if and only if $P \perp P'$ ($\perp$ is the orthogonal relation defined by Ω). We will investigate the pair

$$\Pi_\Omega := (\mathcal{G}_1(\Omega), \mathcal{L}(\Omega)).$$

Since two "proportional" forms (one of the forms is a scalar multiple of the other) have the same set of totally isotropic subspaces, we can assume that Ω is one of the forms given in Theorem 1.6: alternating, symmetric, or Hermitian.

Suppose that

$$\Omega(x, y) = \varepsilon\sigma(\Omega(y, x)) \quad \forall\, x, y \in V,$$

where σ is an anti-automorphism of R satisfying $\sigma^2 = 1_R$ and $\varepsilon = \pm 1$; in other words, Ω is one of the forms considered in Examples 1.9 and 1.10. We define the *trace set*

$$T(\sigma, \varepsilon) := \{\, a + \varepsilon\sigma(a) \,:\, a \in R \,\}$$

and say that the form Ω is *trace-valued* if

$$\Omega(x, x) \in T(\sigma, \varepsilon) \tag{4.2}$$

for every $x \in V$. This condition holds if the characteristic of R is not equal to 2. Also every alternating form is trace-valued.

Remark 4.6. In the general case, all vectors $x \in V$ satisfying (4.2) form a linear subspace $W \subset V$. All isotropic vectors belong to W and the restriction of Ω to W is trace-valued.

Example 4.1. Suppose that R is a field of characteristic 2 and the form Ω is bilinear ($\sigma = 1_R$ and $\varepsilon = 1 = -1$). Since the trace set is zero, the form Ω is trace-valued only in the case when it is alternating.

Lemma 4.6. *Let S be a 2-dimensional linear subspace of V containing an isotropic vector x. If Ω is trace-valued then S contains an isotropic vector linearly independent with x.*

Proof. The statement is trivial if S is totally isotropic. Otherwise, we choose a vector $y \in S$ such that $\Omega(x, y) = 1$ and a scalar $a \in R$ satisfying

$$a + \varepsilon \sigma(a) + \Omega(y, y) = 0$$

(this is possible, since our form is trace-valued). Then

$$\Omega(ax + y, ax + y) = \Omega(ax, ax) + a\Omega(x, y) + \varepsilon \sigma(a\Omega(x, y)) + \Omega(y, y) = 0. \qquad \square$$

In other words, if the form Ω is trace-valued and a line of Π_V has a non-empty intersection with $\mathcal{G}_1(\Omega)$ then this intersection contains more than one point. Using this fact we prove the following.

Lemma 4.7. *If Ω is trace-valued then there is a base of V consisting of isotropic vectors.*

Proof. Let $B = \{x_1, \ldots, x_n\}$ be a base of V and x be an isotropic vector. The intersection of the linear subspaces

$$\langle B \setminus \{x_i\} \rangle, \quad i = 1, \ldots, n,$$

is empty and we choose j such that x does not belong to $\langle B \setminus \{x_j\} \rangle$. Assume that $j = n$. Lemma 4.6 implies the existence of isotropic vectors

$$y_i \in \langle x, x_i \rangle \setminus \{x\}, \quad i = 1, \ldots, n - 1.$$

The base $x, y_1, \ldots, y_{n-1}$ is as required. $\qquad \square$

Proposition 4.12. *If Ω is trace-valued then Π_Ω is a polar space.*

Proof. The polar axiom (1) is trivial and we leave the verification of the axiom (2) for the reader. If the axiom (3) fails then there exists an isotropic vector orthogonal to all isotropic vectors; by Lemma 4.7, this vector is orthogonal to all vectors of V which is impossible, since our form is non-degenerate. The axiom (4) follows from the fact that singular subspaces of Π_Ω can be identified with totally isotropic subspaces. $\qquad \square$

It follows immediately from Theorem 4.1 and Proposition 4.12 that all maximal totally isotropic subspaces have the same dimension. This dimension is called the *Witt index* of the form Ω. The Witt index is not greater than $\frac{n}{2}$ and equal to the rank of the polar space Π_Ω. For example, if Ω is alternating then n is even and the Witt index is equal to $\frac{n}{2}$.

Remark 4.7. The verification of Tits – Veldkamp axioms is more complicated [Dieudonné (1971); Taylor (1992)]. The statement concerning the dimension of maximal totally isotropic spaces is a part of well-known Witt theorem.

Lemma 4.8. *Let σ be a non-identity anti-automorphism of R satisfying $\sigma^2 = 1_R$. Then there exists a non-zero scalar $a \in R$ such that*

$$a + \sigma(a) = 0.$$

Proof. The statement is trivial if the characteristic is equal to 2 (we can take, for example, $a = 1$). Suppose that the characteristic is not equal to 2. In this case, the trace set

$$T(\sigma, 1) = \{\, a + \sigma(a) \ : \ a \in R \,\}$$

coincides with the set of all scalars $b \in R$ satisfying $\sigma(b) = b$. Let us take any $b \in R \setminus T(\sigma, 1)$ (such elements exist, since σ is non-identity). Then

$$c := b + \sigma(b) \in T(\sigma, 1)$$

and the equality

$$c = \frac{c}{2} + \sigma\left(\frac{c}{2}\right)$$

guarantees that the scalar

$$a := b - \frac{c}{2} \neq 0$$

is as required. $\square$

By Theorem 1.6, there are precisely the following three types of polar spaces associated with reflexive forms:

- symplectic polar spaces (defined by alternating forms),
- symmetric polar spaces (defined by symmetric forms, the characteristic is not equal to 2),
- Hermitian polar spaces (defined by trace-valued Hermitian forms).

Proposition 4.13. *Let* Π_Ω *be one of the polar spaces considered above. Then* Π_Ω *is of type* D_m *only in the following case: the characteristic of* R *is not equal to 2, the dimension* n *is even and* Ω *is a symmetric form of Witt index* $\frac{n}{2}$.

Proof. Let m be the Witt index of the form Ω and S be a totally isotropic subspace of dimension $m - 1$. Then S is the intersection of two maximal totally isotropic subspaces U_1 and U_2. We take 1-dimensional linear subspaces P_i $(i = 1, 2)$ such that

$$U_i = S + P_i.$$

It is clear that $P_1 \not\perp P_2$.

Suppose that Ω is alternating. Then for every 1-dimensional linear subspace $P \subset P_1 + P_2$ the linear subspace $S + P$ is totally isotropic and our polar space is of type C_m.

Consider the case when Ω is a Hermitian form associated with an anti-automorphism $\sigma : R \to R$. If $x, y \in V$ are isotropic vectors then

$$\Omega(x + by, x + by) = \Omega(x, by) + \Omega(by, x) = \sigma(b\Omega(y, x)) + b\Omega(y, x).$$

We choose vectors $x \in P_1$, $y \in P_2$ and a scalar $b \in R$ such that $a = b\Omega(y, x)$ satisfies the condition of Lemma 4.8. Then $\langle x + by \rangle$ belongs to $\mathcal{G}_1(\Omega)$ and

$$S + \langle x + by \rangle$$

is a third maximal totally isotropic subspace containing S. The polar space is of type C_m.

Now suppose that Ω is symmetric and the characteristic of R is not equal to 2. We have

$$\Omega(x + by, x + by) = 2b\Omega(x, y)$$

for any isotropic vectors x and y. Since $P_1 \not\perp P_2$, this means that the linear subspace $P_1 + P_2$ does not contain elements of $\mathcal{G}_1(\Omega)$ distinct from P_1 and P_2.

If $m < \frac{n}{2}$ then $S + P_1 + P_2$ is a proper linear subspace of $S^\perp$, since

$$\dim(S + P_1 + P_2) = m + 1 < n - m + 1 = \dim S^\perp.$$

We take any 1-dimensional linear subspace $P' \subset S^\perp$ which is not contained in $S + P_1 + P_2$. By Lemma 4.6, $P_1 + P'$ contains $P'' \in \mathcal{G}_1(\Omega)$ distinct from P_1. Then $S + P''$ is a third maximal totally isotropic subspace containing S. The polar space is of type C_m.

Every maximal totally isotropic subspace U containing S is contained in $S^\perp$. If $n = 2m$ then $S^\perp$ is $(m + 1)$-dimensional and U intersects $P_1 + P_2$ in a 1-dimensional linear subspace P. Since P is totally isotropic, it coincides with P_1 or P_2. This implies that there are precisely two maximal totally isotropic subspaces containing S and the polar space is of type D_m. $\quad\square$

Frames of the polar space Π_Ω are independent subsets of Π_V. If the Witt index is less than $\frac{n}{2}$ then every frame spans a proper subspace of Π_V whose intersection with $\mathcal{G}_1(\Omega)$ is a subspace of Π_Ω; this subspace is proper (by Lemma 4.7) and frames are not bases of Π_Ω. However, if the characteristic of R is not equal to 2 and Ω is alternating then every frame is a base of Π_Ω [Blok and Brouwer (1998); Cooperstein and Shult (1997)]. In the case of characteristic 2, this fails.

Example 4.2. Suppose that $n = 4$, R is a field of characteristic 2 and the form Ω is alternating. Let $B = \{p_1, p_2, q_1, q_2\}$ be a frame of the polar space Π_Ω such that $p_i \not\perp q_i$ for each i. There exist vectors $x_1, x_2, y_1, y_2 \in V$ satisfying

$$p_i = \langle x_i \rangle, \ q_i = \langle y_i \rangle \ \text{and} \ \Omega(x_i, y_i) = \Omega(y_i, x_i) = 1.$$

Consider two pairs of collinear points

$$p' \in p_1 p_2, \ q' \in q_1 q_2 \ \text{and} \ p'' \in p_1 q_2, \ q'' \in p_2 q_1$$

distinct from the points of B. Then for some non-zero scalars $a, b \in R$ we have

$$p' = \langle x_1 + a x_2 \rangle, \ q' = \langle y_2 + a y_1 \rangle,$$

$$p'' = \langle x_1 + b y_2 \rangle, \ q'' = \langle x_2 + b y_1 \rangle.$$

Almost all points of the line $p' q'$ (except the point q') are of type

$$\langle (x_1 + a x_2) + t(y_2 + a y_1) \rangle, \ t \in R,$$

and almost all points of the line $p'' q''$ (except the point q'') are of type

$$\langle (x_1 + b y_2) + s(x_2 + b y_1) \rangle, \ s \in R.$$

These lines have a common point ($t = b, s = a$). This means that $\langle B \rangle$ is the grid consisting of all lines which intersect $p_1 p_2$ and $q_1 q_2$ or $p_1 q_2$ and $p_2 q_1$. Therefore, $\langle B \rangle$ is a proper subspace of Π_Ω; it does not contain, for example, any line passing through p_1 and distinct from $p_1 p_2$ and $p_1 q_2$.

4.3.2 *Polar spaces associated with quadratic forms*

Let V be an n-dimensional vector space over a field F. A non-zero mapping $Q : V \to F$ is called a *quadratic form* if

$$Q(ax) = a^2 Q(x)$$

for all $a \in F, x \in V$ and there exists a bilinear form $\Omega : V \times V \to F$ such that

$$Q(x + y) = Q(x) + Q(y) + \Omega(x, y)$$

for all $x, y \in V$. It is clear that Ω is symmetric. The form Q is said to be *non-degenerate* if $Q(x) \neq 0$ for every non-zero vector x belonging to $V^\perp$ (we do not require that the associated form Ω is non-degenerate and write $V^\perp$ for the linear subspace of all vectors orthogonal to V). By this definition, Q is non-degenerate if the associated form Ω is non-degenerate.

A non-zero vector $x \in V$ is called *singular* if $Q(x) = 0$ and we say that a linear subspace S is *totally singular* if every non-zero vector of S is singular. Every totally singular subspace is totally isotropic for the associated form Ω; but totally isotropic subspaces of Ω need not to be totally singular. If the characteristic of F is not equal to 2 then Q can be uniquely recovered from Ω by the formula

$$Q(x) = \frac{\Omega(x, x)}{2}.$$

This equality guarantees that Q is non-degenerate if and only if Ω is non-degenerate; moreover, a linear subspace is totally singular (for Q) if and only if it is totally isotropic (for Ω).

From this moment we assume that Q is non-degenerate and there exist totally singular subspaces of dimension 2. We write $\mathcal{G}_1(Q)$ for the set of all 1-dimensional totally singular subspaces and denote by $\mathcal{L}(Q)$ the set formed by all lines of Π_V such that the associated 2-dimensional linear subspaces are totally singular. Consider the pair

$$\Pi_Q := (\mathcal{G}_1(Q), \mathcal{L}(Q)).$$

It is clear that $P, P' \in \mathcal{G}_1(Q)$ are joined in Π_Q by a line only in the case when $P \perp P'$ ($\perp$ is the orthogonal relation defined by Ω). If the characteristic of F is not equal to 2 then Π_Q coincides with the polar space Π_Ω.

Suppose that the characteristic of F is equal to 2. The equality

$$Q(2x) = 2Q(x) + \Omega(x, x)$$

shows that Ω is alternating. We restrict our self to the case when Ω is non-degenerate and refer [Dieudonné (1971)] for the general case. Since totally singular subspaces are totally isotropic, Π_Q is a subspace of the polar space Π_Ω. Thus Π_Q satisfies the polar axioms (1), (2) and (4). This is a proper subspace of Π_Ω (all vectors are isotropic, but $Q \not\equiv 0$ and there are non-singular vectors).

Exercise 4.4. Prove the following analogue of Lemma 4.6: for every singular vector $x \in V$ and any vector $y \in V$ such that $\Omega(x, y) \neq 0$ there exists a singular vector $z \in \langle x, y \rangle$ independent with x. *Hint:* take a scalar $a \in F$ satisfying $a\Omega(x, y) + Q(y) = 0$, the vector $ax + y$ is as required.

Exercise 4.5. Prove the analogue of Lemma 4.7: there exists a base of V consisting of singular vectors. *Hint:* apply Exercise 4.4 to a singular vector $x \in V$ and a base of V formed by vectors non-orthogonal to x.

Using Exercise 4.5 we establish that Π_Q satisfies the axiom (3). Therefore, Π_Q is a polar space. All singular subspaces of Π_Q can be identified with totally singular subspaces of Q.

We do not consider here so-called *pseudo-quadratic* forms associated with Hermitian forms (see [Tits (1974)], Section 8.2); as quadratic forms, they give new examples of polar spaces only in the case of characteristic 2.

4.3.3 *Polar spaces of type* D$_3$

Let V be a 4-dimensional vector space over a division ring R.

Exercise 4.6. Show that the Grassmann space $\mathfrak{G}_2(V)$ is a polar space.

Maximal singular subspaces of $\mathfrak{G}_2(V)$ (stars and tops) are projective planes, thus the rank of the polar space is equal to 3. Since every line is contained in precisely one star and precisely one top, this is a polar space of type D$_3$.

Remark 4.8. If R is commutative then this polar space can be obtained from the Klein quadratic form defined on the 6-dimensional vector space $\wedge^2 V$ [Cameron (1991); Taylor (1992)].

Conversely, let $\Pi = (P, \mathcal{L})$ be a polar space of type D$_3$. Consider the associated half-spin Grassmannian space $\mathfrak{G}_\delta(\Pi)$, $\delta \in \{+, -\}$. There is natural one-to-one correspondence between lines of $\mathfrak{G}_\delta(\Pi)$ and points of Π (every line of $\mathfrak{G}_\delta(\Pi)$ consists of all elements of $\mathcal{G}_\delta(\Pi)$ passing through a certain point). Any two distinct elements of $\mathcal{G}_\delta(\Pi)$ are adjacent and $\mathfrak{G}_\delta(\Pi)$ is a linear space.

Lemma 4.9. *If* Π *is a polar space of type* D_3 *then* $\mathfrak{G}_\delta(\Pi)$, $\delta \in \{+, -\}$, *is a 3-dimensional projective space.*

Proof.   Let S_1, S_2, S_3 be a triangle in $\mathfrak{G}_\delta(\Pi)$. Then

$$p_1 = S_2 \cap S_3, \; p_2 = S_1 \cap S_3, \; p_3 = S_1 \cap S_2$$

form a triangle in Π and span a certain plane U. This plane belongs to $\mathcal{G}_{-\delta}(\Pi)$, since each S_i intersects U in the line $p_l p_m$, where $i \neq l, m$.

Denote by $\mathcal{X}$ the set consisting of all elements of $\mathcal{G}_\delta(\Pi)$ which intersect U in lines (for every $S \in \mathcal{G}_\delta(\Pi)$ the intersection $S \cap U$ is empty or a line). If S and S' are distinct elements of $\mathcal{X}$ then $S \cap S' \cap U$ is a certain point p and the associated line $[p\rangle_\delta$ joins S and S'; it is clear that this line is contained in $\mathcal{X}$. Therefore, $\mathcal{X}$ is a subspace of $\mathfrak{G}_\delta(\Pi)$ and a line of $\mathfrak{G}_\delta(\Pi)$ is contained in $\mathcal{X}$ if and only if the associated point belongs to U.

There is a one-to-one correspondence between elements of $\mathcal{X}$ and lines of U (every line of U is contained in precisely one element of $\mathcal{X}$). This is a collineation of the subspace $\mathcal{X}$ to the projective plane dual to U. Clearly, S_1, S_2, S_3 form a base of the projective plane $\mathcal{X}$ and we have established that every plane of $\mathfrak{G}_\delta(\Pi)$ is projective.

If $p \in P \setminus U$ then $U \cap p^\perp$ is a line in U and the maximal singular subspace $\langle U \cap p^\perp, p \rangle$ belongs to $\mathcal{G}_\delta(\Pi)$; in other words, the line of $\mathfrak{G}_\delta(\Pi)$ defined by the point p has a non-empty intersection with $\mathcal{X}$. This means that $\mathcal{X}$ is a hyperplane of $\mathfrak{G}_\delta(\Pi)$ and we get the claim. $\qquad\square$

So, we can suppose that $\mathfrak{G}_+(\Pi) = \Pi_V$, where V is a 4-dimensional vector space over a division ring. Consider the bijection

$$f : \mathcal{G}_2(V) \to P$$

induced by the natural one-to-one correspondence between lines of $\mathfrak{G}_+(\Pi)$ and points of Π. It was established above that for every plane $\mathcal{X}$ of $\mathfrak{G}_+(\Pi)$ there exists $U \in \mathcal{G}_-(\Pi)$ such that $\mathcal{X}$ consists of all elements of $\mathcal{G}_+(\Pi)$ intersecting U in lines. This is a one-to-one correspondence and we get another bijection

$$g : \mathcal{G}_3(V) \to \mathcal{G}_-(\Pi).$$

Exercise 4.7. Show that f is a collineation of $\mathfrak{G}_2(V)$ to Π and g is a collineation of Π_V^* to $\mathfrak{G}_-(\Pi)$.

Exercise 4.8. Let h be a collineation of $\mathfrak{G}_2(V)$ to Π. Show that $\mathcal{A} \subset \mathcal{G}_2(V)$ is an apartment if and only if $h(\mathcal{A})$ is a frame of Π. *Hint:* use Theorem 3.8.

By Exercises 4.7 and 4.8, we get the following.

Proposition 4.14. *If Π is a polar space of type D_3 then there exists a 4-dimensional vector space V such that*

$$\Pi,\ \mathfrak{G}_+(\Pi),\ \mathfrak{G}_-(\Pi)$$

are isomorphic to

$$\mathfrak{G}_2(V),\ \Pi_V,\ \Pi_V^*,$$

respectively. Moreover, every collineation of Π to $\mathfrak{G}_2(V)$ establishes a one-to-one correspondence between frames of Π and apartments of $\mathcal{G}_2(V)$.

4.3.4 *Embeddings in projective spaces and classification*

The *Veldkamp space* $\mathfrak{V}(\Pi)$ of a polar space Π (the rank of Π is assumed to be not less than 3) is the linear space whose points are hyperplanes of Π and the line joining hyperplanes H_1 and H_2 consists of all hyperplanes containing $H_1 \cap H_2$. The mapping $p \to p^\perp$ is an embedding of Π in $\mathfrak{V}(\Pi)$ and it is not difficult to prove that this embedding transfers lines to lines.

Theorem 4.5 ([Veldkamp (1959/1960)]). *Let Π be a polar space of rank $n \geq 3$; in the case when $n = 3$, we require in addition that Π is of type C_n and every 2-dimensional singular subspace is a Desarguesian projective plane. Then $\mathfrak{V}(\Pi)$ is a projective space.*

This result was exploited in the following classification of polar spaces.

Theorem 4.6 ([Tits (1974)]). *Up to isomorphism there are the following three types of polar spaces whose rank is not less than three:*

- *the polar spaces associated with non-degenerate reflexive sesquilinear, quadratic, and pseudo-quadratic forms;*
- *the Grassmann space $\mathfrak{G}_2(V)$, where V is a 4-dimensional vector space, this polar space is defined by Klein's quadratic form if V is a vector space over a field;*
- *the polar spaces of type C_3 associated with Cayley algebras, maximal singular subspaces of such polar spaces are non-Desarguesian (Moufang) projective planes.*

In particular, every polar space whose rank is greater than 3 is isomorphic to the polar space associated with a certain reflexive sesquilinear, quadratic, or pseudo-quadratic form.

4.4 Polar buildings

4.4.1 *Buildings of type* C_n

Let $\Pi = (P, \mathcal{L})$ be a polar space of rank $n \geq 3$. Consider the flag complex $\Delta(\Pi)$ obtained from the set of all proper singular subspaces of Π with the natural incidence relation. Every frame B of Π defines the subcomplex $\Sigma_B \subset \Delta(\Pi)$ consisting of all flags formed by singular subspaces spanned by subsets of B; this subcomplex is called the *apartment* associated with (defined by) the frame B. The complex $\Delta(\Pi)$ together with the set of all such apartments is a building of type C_n. This building is thick only in the case when Π is of type C_n. The Grassmannians of $\Delta(\Pi)$ are the polar Grassmannians $\mathcal{G}_k(\Pi)$, $k \in \{0, 1, \ldots, n-1\}$; the corresponding Grassmann spaces are $\mathfrak{G}_k(\Pi)$. In what follows the associated Grassmann graphs (the collinearity graphs of Grassmann spaces) will be denoted by $\Gamma_k(\Pi)$.

By [Tits (1974)], every thick building of type C_n ($n \geq 3$) is isomorphic to the C_n-building of a rank n polar space.

4.4.2 *Buildings of type* D_n

Now let $\Pi = (P, \mathcal{L})$ be a polar space of type D_n and $n \geq 4$. We write $\mathcal{G}^*(\Pi)$ for the set of all proper singular subspaces whose dimension is not equal to $n - 2$ and define the *oriflamme* incidence relation $*$ on $\mathcal{G}^*(\Pi)$:

- if $S \in \mathcal{G}_k(\Pi)$, $k \leq n - 3$, and $U \in \mathcal{G}^*(\Pi)$ then $S * U$ means that S and U are incident in the usual sense,
- for $S \in \mathcal{G}_+(\Pi)$ and $U \in \mathcal{G}_-(\Pi)$ we write $S * U$ if their intersection belongs to $\mathcal{G}_{n-2}(\Pi)$.

The associated flag complex will be called the *oriflamme* complex of Π and denoted by $\mathrm{Orif}(\Pi)$. For every frame B the corresponding *apartment* Orif_B is the subcomplex of $\mathrm{Orif}(\Pi)$ consisting of all oriflamme flags formed by singular subspaces spanned by subsets of B. The complex $\mathrm{Orif}(\Pi)$ together with the set of all such apartments is a thick building of type D_n. The Grassmannians are the polar Grassmannians $\mathcal{G}_k(\Pi)$, $k \in \{0, 1, \ldots, n-3\}$, and the half-spin Grassmannians $\mathcal{G}_\delta(\Pi)$, $\delta \in \{+, -\}$. The corresponding Grassmann spaces are $\mathfrak{G}_k(\Pi)$ and $\mathfrak{G}_\delta(\Pi)$; we write $\Gamma_k(\Pi)$ and $\Gamma_\delta(\Pi)$ for the associated Grassmann graphs.

Every thick building of type D_n ($n \geq 4$) is isomorphic to the oriflamme building obtained from a polar space of type D_n [Tits (1974)].

4.5 Elementary properties of Grassmann spaces

4.5.1 *Polar Grassmann spaces*

Let $\Pi = (P, \mathcal{L})$ be a polar space of rank $n \geq 3$. First, we describe triangles in the Grassmann space $\mathfrak{G}_k(\Pi)$, $1 \leq k \leq n - 1$.

Lemma 4.10. *If $1 \leq k \leq n - 2$ and $S_1, S_2, S_3 \in \mathcal{G}_k(\Pi)$ form a triangle in $\mathfrak{G}_k(\Pi)$ then $S_i \perp S_j$ for all $i, j \in \{1, 2, 3\}$ and only one of the following possibilities is realized:*

(1) *a star-triangle: $k < n - 2$ and there is a $(k-1)$-dimensional singular subspace contained in each S_i,*
(2) *a top-triangle: there is a $(k+1)$-dimensional singular subspace containing all S_i.*

The Grassmann space $\mathfrak{G}_{n-1}(\Pi)$ does not contain triangles: any three mutually collinear points of $\mathfrak{G}_{n-1}(\Pi)$ are collinear if Π is of type C_n and $\mathfrak{G}_{n-1}(\Pi)$ does not contain triples of mutually collinear points if Π is of type D_n.

Proof. Let $1 \leq k \leq n - 2$ and $S_1, S_2, S_3 \in \mathcal{G}_k(\Pi)$ be a triangle in $\mathfrak{G}_k(\Pi)$. Since S_i is adjacent with S_j ($i \neq j$), we have $S_i \perp S_j$ for all $i, j \in \{1, 2, 3\}$. As for Grassmannians of finite-dimensional vector spaces, if S_3 is not contained in the $(k+1)$-dimensional singular subspace spanned by S_1 and S_2 then

$$\langle S_1, S_2 \rangle \cap S_3$$

is a $(k-1)$-dimensional singular subspace contained in S_1 and S_2; in this case, the singular subspace spanned by S_1, S_2, S_3 is $(k+2)$-dimensional which is possible only for $k < n - 2$.

Let $S_1, S_2, S_3 \in \mathcal{G}_{n-1}(\Pi)$ be mutually collinear points of $\mathfrak{G}_{n-1}(\Pi)$. Then

$$U_1 := S_2 \cap S_3, \ U_2 := S_1 \cap S_3, \ U_3 := S_1 \cap S_2$$

belong to $\mathcal{G}_{n-2}(\Pi)$. Since

$$U_1, U_2 \subset S_3, \ U_1, U_3 \subset S_2, \ U_2, U_3 \subset S_1,$$

we have $U_i \perp U_j$ for all $i, j \in \{1, 2, 3\}$. Suppose that at least two of these three subspaces are distinct, for example, $U_1 \neq U_2$. Then

$$S_3 = \langle U_1, U_2 \rangle$$

and $U_3 \perp S_3$. The singular subspace S_3 is maximal and we get the inclusion

$$S_1 \cap S_2 = U_3 \subset S_3$$

which implies that U_3 is contained in each S_i. Hence $U_1 = U_2 = U_3$, a contradiction. Therefore, all U_i are coincident and S_1, S_2, S_3 are points on a line of $\mathfrak{G}_{n-1}(\Pi)$. $\qquad\square$

Proposition 4.15. $\mathfrak{G}_k(\Pi)$ *is a gamma space.*

Proof. By Lemma 4.10, if a point of $\mathfrak{G}_{n-1}(\Pi)$ is collinear with two distinct points on a line then it belongs to this line and $\mathfrak{G}_{n-1}(\Pi)$ is a gamma space. The case $k = 0$ was considered in Subsection 4.1.1 and we suppose that $0 < k < n-1$. If $S \in \mathcal{G}_k(\Pi)$ is collinear with two distinct points S_1, S_2 of a line $[M, N]_k$ and S does not belong to this line then $S \perp N$ (since N is spanned by S_1 and S_2) and

$$M = S_1 \cap S_2 \subset S \text{ or } S \subset N$$

$(S, S_1, S_2$ form a star-triangle or a top-triangle, respectively). In each of these cases, S is collinear with all points of the line $[M, N]_k$. $\qquad\square$

It follows from Proposition 4.15 that the class of maximal singular subspaces of $\mathfrak{G}_k(\Pi)$ coincides with the class of maximal cliques of the Grassmann graph $\Gamma_k(\Pi)$.

Example 4.3. By the second part of Lemma 4.10, every line of $\mathfrak{G}_{n-1}(\Pi)$ is a maximal singular subspace.

Example 4.4. Suppose that $k \leq n-2$. For every $(k+1)$-dimensional singular subspace N the set $\langle N]_k$ is called a *top*. This is a singular subspace of $\mathfrak{G}_k(\Pi)$ isomorphic to a $(k+1)$-dimensional projective space. As in Example 3.1, we show that this singular subspace is maximal. Every triangle in a top is a top-triangle.

Example 4.5. As in the previous example, we suppose that $k \leq n-2$. Let M be a $(k-1)$-dimensional singular subspace and N be a maximal singular subspace containing M. Then $[M, N]_k$ is a singular subspace of $\mathfrak{G}_k(\Pi)$. In the case when $k = n-2$, this is a line. If $k < n-2$ then the subspace $[M, N]_k$ is said to be a *star*. It can be identified with a star in the Grassmann space $\langle N]_k$. The star $[M, N]_k$ is isomorphic to an $(n-k-1)$-dimensional projective space. Let S be a k-dimensional singular subspace of Π which does not belong to $[M, N]_k$. If $S \subset N$ then there exists an element of $[M, N]_k$ non-adjacent with S (since $[M, N]_k$ is a maximal clique in the collinearity graph of $\langle N]_k$). If S does not belong to $\langle N]_k$ then $S \not\perp N$

and we take any point $p \in N$ satisfying $p \not\perp S$; every element of $[M, N]_k$ containing p (it is clear that such elements exist) is non-adjacent with S. Therefore, $[M, N]_k$ is a maximal singular subspace of $\mathfrak{G}_k(\Pi)$. Every triangle in a star is a star-triangle.

Proposition 4.16. *The following assertions are fulfilled:*

(1) *every maximal singular subspace of $\mathfrak{G}_{n-1}(\Pi)$ is a line;*
(2) *if $k \leq n - 2$ then every maximal singular subspace of $\mathfrak{G}_k(\Pi)$ is a top or a star; in particular, all maximal singular subspaces of $\mathfrak{G}_{n-2}(\Pi)$ are tops.*

Proof. It was noted above that the class of maximal singular subspaces of $\mathfrak{G}_k(\Pi)$ coincides with the class of maximal cliques of the Grassmann graph $\Gamma_k(\Pi)$. The statement (1) follows from the second part of Lemma 4.10.

(2). Let $\mathcal{X}$ be a maximal clique of $\Gamma_k(\Pi)$ and $k \leq n-2$. Suppose that $\mathcal{X}$ is not a star. As in the proof of Proposition 3.2, we choose $S_1, S_2, S_3 \in \mathcal{X}$ which form a top-triangle in $\mathfrak{G}_k(\Pi)$. If $S \in \mathcal{G}_k(\Pi)$ is not contained in the $(k + 1)$-dimensional singular subspace

$$N := \langle S_1, S_2 \rangle$$

then S is non-adjacent with at least one of S_i. This mean that $S \notin \mathcal{X}$ and our clique coincides with the top $\langle N]_k$. $\square$

Example 4.6. Let M be an m-dimensional singular subspace and $m < k$. Then $[M\rangle_k$ is a subspace of $\mathfrak{G}_k(\Pi)$; subspaces of such type are called *parabolic* [Cooperstein, Kasikova and Shult (2005)]. This subspace is singular only in the case when $k = n - 1$ and $m = n - 2$. Suppose that $m < n - 2$. By Lemma 4.4, $[M\rangle_{m+1}$ is a polar space of rank $n - m - 1$. The parabolic subspace $[M\rangle_k$ can be identified with the Grassmann space of index $k - m - 1$ associated with this polar space (every $U \in [M\rangle_k$ corresponds to $[M, U]_{m+1}$).

Example 4.7. Let $n \geq 4$. If $0 < k < n - 1$ and M, N is a pair of incident singular subspaces of Π such that

$$\dim M < k < \dim N$$

then $[M, N]_k$ is a subspace of $\mathfrak{G}_k(\Pi)$ isomorphic to the Grassmann space of a finite-dimensional vector space (in the case when $\dim M = k - 1$ and $\dim N = k + 1$, we get a line). Subspaces of such type are called *classical* [Cooperstein (2005)].

Let S and U be adjacent elements of $\mathcal{G}_k(\Pi)$. The set of all elements of $\mathcal{G}_k(\Pi)$ adjacent with both S and U is the union of all maximal cliques of $\Gamma_k(\Pi)$ (maximal singular subspaces of $\mathfrak{G}_k(\Pi)$) containing S and U. In the case when $k \neq n - 2$, the intersection of all these cliques is the line joining S and U. For $k = n - 2$ this fails, since there is only one maximal clique (the top $\langle\langle S, U \rangle]_{n-2}$) containing S and U.

For every subset $\mathcal{X} \subset \mathcal{G}_k(\Pi)$ we denote by $\mathcal{X}^\sim$ the set consisting of all elements of $\mathcal{G}_k(\Pi)$ adjacent with every element of $\mathcal{X}$. As for the Grassmann spaces of finite-dimensional vector spaces, we have the following characterization of lines in terms of the adjacency relation if $k \neq n - 2$.

Proposition 4.17. *Let $k \neq n - 2$. Then for any adjacent $S, U \in \mathcal{G}_k(\Pi)$ the line joining S and U coincides with the set $\{S, U\}^{\sim\sim}$.*

4.5.2   Half-spin Grassmann spaces

Let $\Pi = (P, \mathcal{L})$ be a polar space of type D_n. We investigate the associated half-spin Grassmann spaces $\mathfrak{G}_\delta(\Pi)$, $\delta \in \{+, -\}$ (they are defined only for $n \geq 3$). If $n = 3$ then both $\mathfrak{G}_\delta(\Pi)$ are 3-dimensional projective spaces (Subsection 4.3.3) and we will suppose that $n \geq 4$.

Lemma 4.11. *Let S_1, S_2, S_3 be a triangle in $\mathfrak{G}_\delta(\Pi)$. Then there exist unique $M \in \mathcal{G}_{n-4}(\Pi)$ and $U \in \mathcal{G}_{-\delta}(\Pi)$ such that $M \subset U$,*
$$M = S_1 \cap S_2 \cap S_3$$
and each S_i has an $(n - 2)$-dimensional intersection with U.

Proof.   The subspaces
$$U_1 := S_2 \cap S_3, \ U_2 := S_1 \cap S_3, \ U_3 := S_1 \cap S_2$$
belong to $\mathcal{G}_{n-3}(\Pi)$. We have $U_i \perp U_j$ for all $i, j \in \{1, 2, 3\}$ (since U_i and U_j both are contained in S_l with $l \neq i, j$). The equality $U_l = U_j$ $(l \neq j)$ implies that U_l is contained in all S_i and we get $U_1 = U_2 = U_3$ which means that S_1, S_2, S_3 are points on a line, a contradiction. Therefore, all U_i are distinct and the dimension of the singular subspace
$$U := \langle U_1, U_2, U_3 \rangle$$
is not less than $n - 2$.

Suppose that $\dim U = n - 2$ or two distinct U_i, for example, U_1 and U_2 are non-adjacent. In each of these cases, the subspace U is spanned by U_1 and U_2. Then $U \subset S_3$ (since U_1 and U_2 both are contained in S_3) and
$$S_1 \cap S_2 = U_3 \subset U \subset S_3;$$
in other words, U_3 is contained in all S_i which is impossible.

So, U is a maximal singular subspace and U_1, U_2, U_3 are mutually adjacent. This means that U_1, U_2, U_3 form a star-triangle in $\mathfrak{G}_{n-3}(\Pi)$. Then the $(n-4)$-dimensional singular subspace

$$M := U_1 \cap U_2 \cap U_3$$

coincides with $S_1 \cap S_2 \cap S_3$. Each S_i intersects U in the $(n-2)$-dimensional subspace $\langle U_l, U_m \rangle$, where $l, m \neq i$. Hence U belongs to $\mathcal{G}_{-\delta}(\Pi)$.

Let U' be a maximal singular subspace of Π intersecting each S_i in an $(n-2)$-dimensional subspace. Since Π is of type D_n, $U' \cap S_i$ and $U' \cap S_j$ $(i \neq j)$ are distinct and their intersection is $(n-3)$-dimensional. On the other hand, this intersection is contained in $S_i \cap S_j = U_l$ $(l \neq i, j)$ which implies that it coincides with U_l. Therefore, each U_l is contained in U' and $U = U'$. $\qquad\square$

Proposition 4.18. $\mathfrak{G}_\delta(\Pi)$ *is a gamma space.*

Proof. Let S_1, S_2, S_3 be a triangle in $\mathfrak{G}_\delta(\Pi)$. We need to show that S_1 is collinear with all points of the line joining S_2 and S_3. If S is a point on this line then $S_2 \cap S_3 \subset S$; in particular, S contains the $(n-4)$-dimensional singular subspace

$$S_1 \cap S_2 \cap S_3$$

(Lemma 4.11). This guarantees that $S_1 \cap S$ is $(n-3)$-dimensional. $\qquad\square$

We want to describe maximal singular subspaces of $\mathfrak{G}_\delta(\Pi)$. By Proposition 4.18, the class of maximal singular subspaces coincides with the class of maximal cliques of the associated Grassmann graph $\Gamma_\delta(\Pi)$.

Let $U \in \mathcal{G}_{-\delta}(\Pi)$. We write $[U]_\delta$ for the set of all elements of $\mathcal{G}_\delta(\Pi)$ intersecting U in $(n-2)$-dimensional subspaces.

Proposition 4.19. *For every $U \in \mathcal{G}_{-\delta}(\Pi)$ the set $[U]_\delta$ is a maximal singular subspace of $\mathfrak{G}_\delta(\Pi)$ isomorphic to an $(n-1)$-dimensional projective space.*

Singular subspaces of such kind will be called *special*.

Proof. As in the proof of Lemma 4.9, we establish that $[U]_\delta$ is a singular subspace of $\mathfrak{G}_\delta(\Pi)$ and a line $[T]_\delta$, $T \in \mathcal{G}_{n-3}(\Pi)$, is contained in $[U]_\delta$ if and only if $T \subset U$. Every hyperplane of U is contained in precisely one element of $[U]_\delta$; this correspondence defines a collineation between $[U]_\delta$ and the projective space dual to U. So, $[U]_\delta$ is an $(n-1)$-dimensional projective space.

For every $S \in \mathcal{G}_\delta(\Pi) \setminus [U]_\delta$ there exists a frame B such that S and U are spanned by subsets of B (Proposition 4.7). It is not difficult to construct

$$\langle X \rangle \in [U]_\delta, \quad X \subset B,$$

non-adjacent with S (we leave the details for the reader). This means that $[U]_\delta$ is a maximal singular subspace of $\mathfrak{G}_\delta(\Pi)$. $\qquad\square$

Proposition 4.20. *If $M \in \mathcal{G}_{n-4}(\Pi)$ then $[M\rangle_\delta$ is a maximal singular subspace of $\mathfrak{G}_\delta(\Pi)$ isomorphic to a 3-dimensional projective space.*

Maximal singular subspace of such kind are said to be *stars*.

Proof. If S and U are distinct elements of $[M\rangle_\delta$ then they both contain the $(n-4)$-dimensional singular subspace M which means that their intersection is $(n-3)$-dimensional and S, U are adjacent; moreover, the inclusion $M \subset S \cap U$ guarantees that the line $[S \cap U\rangle_\delta$ joining S and U is contained in $[M\rangle_\delta$. Hence $[M\rangle_\delta$ is a singular subspace of $\mathfrak{G}_\delta(\Pi)$.

Let S_1, S_2, S_3 be a triangle contained in $[M\rangle_\delta$ and U be the associated element of $\mathcal{G}_{-\delta}(\Pi)$ (see Lemma 4.11). Consider the subspace

$$\mathcal{X} := [U]_\delta \cap [M\rangle_\delta.$$

If $[M_i\rangle_\delta$ $(i = 1, 2)$ are distinct lines of $\mathcal{X}$ then

$$M_1 \cap M_2 = M \quad \text{and} \quad M_i \subset U, \quad i = 1, 2;$$

in other words, M_1, M_2 are adjacent elements of $\mathcal{G}_{n-3}(\Pi)$ and the subspace spanned by them is $(n-2)$-dimensional. There is a unique element of $\mathcal{G}_\delta(\Pi)$ intersecting U in $\langle M_1, M_2 \rangle$; it is a common point of our lines. Thus $\mathcal{X}$ is a projective plane and S_1, S_2, S_3 form a base of $\mathcal{X}$. We have proved that every plane in $[M\rangle_\delta$ is projective.

Suppose that $T \in \mathcal{G}_{n-3}(\Pi)$ contains M and $T \not\subset U$. Then $[T\rangle_\delta$ is a line of $[M\rangle_\delta$ which is not contained in $\mathcal{X}$. The maximal singular subspace

$$\langle T, U \cap T^\perp \rangle$$

intersects U in the $(n-2)$-dimensional subspace $U \cap T^\perp$ (by Lemma 4.2); hence it belongs to $\mathcal{X}$ and the line $[T\rangle_\delta$ has a non-empty intersection with $\mathcal{X}$. This means that the plane $\mathcal{X}$ is a hyperplane of $[M\rangle_\delta$. Therefore, $[M\rangle_\delta$ is a 3-dimensional projective space.

Let $S \in \mathcal{G}_\delta(\Pi) \setminus [M\rangle_\delta$. As in the proof of Proposition 4.19, we take a frame B such that S and M are spanned by subsets of B and construct

$$\langle X \rangle \in [M\rangle_\delta, \quad X \subset B,$$

non-adjacent with S. This guarantees that $[M\rangle_\delta$ is a maximal singular subspace of $\mathfrak{G}_\delta(\Pi)$. □

Proposition 4.21. *Every maximal singular subspace of $\mathfrak{G}_\delta(\Pi)$ is a star or a special subspace.*

Proof. We show that every maximal clique $\mathcal{X}$ of the Grassmann graph $\Gamma_\delta(\Pi)$ is a star or a special subspace. Lines are not maximal cliques of $\Gamma_\delta(\Pi)$ (they are contained, for example, in stars), thus $\mathcal{X}$ contains a triangle S_1, S_2, S_3. Let M and U be the associated elements of $\mathcal{G}_{n-4}(\Pi)$ and $\mathcal{G}_{-\delta}(\Pi)$, respectively (Lemma 4.11). Clearly, $[M\rangle_\delta$ and $[U]_\delta$ are the unique star and the unique special subspace containing our triangle.

First, we establish that every $S \in \mathcal{G}_\delta(\Pi)$ adjacent with all S_i belongs to $[M\rangle_\delta$ or $[U]_\delta$. Since $U \cap S_i$ is a hyperplane of S_i, the subspace S intersects each $U \cap S_i$ in a subspace whose dimension is not less than $n - 4$. One of the following two possibilities is realized:

- $S \cap U \cap S_1 = S \cap U \cap S_2 = S \cap U \cap S_3$,
- $S \cap U \cap S_i \neq S \cap U \cap S_j$ for some i, j.

In the first case, the equality

$$M = \bigcap_{i=1}^{3} (U \cap S_i)$$

guarantees that

$$S \cap U \cap S_i = S \cap M$$

for each i. Since M is $(n-4)$-dimensional and the dimension of $S \cap U \cap S_i$ is not less than $n-4$, we have $M \subset S$ and S belongs to $[M\rangle_\delta$. In the second case, the dimension of $S \cap U$ is not less than $n - 3$; thus $S \cap U$ is $(n-2)$-dimensional and $S \in [U]_\delta$.

Now consider

$$S \in [M\rangle_\delta \setminus [U]_\delta \quad \text{and} \quad S' \in [U]_\delta \setminus [M\rangle_\delta.$$

Then $S \cap U$ and $S' \cap U$ span U (the first subspace contains M and the second is a hyperplane of U which does not contain M). Hence every point of $S \cap S'$ is collinear with all points of U and we get

$$S \cap S' \subset U$$

(since U is a maximal singular subspace). The latter inclusion implies that

$$S \cap S' = (S \cap U) \cap (S' \cap U).$$

Since $S \cap U = M$ (this follows from $S \in [M\rangle_\delta \setminus [U]_\delta$) and $S' \cap U$ does not contain M, the subspaces S and S' are not adjacent.

So, every element of $[M\rangle_\delta \setminus [U]_\delta$ is non-adjacent with every element of $[U]_\delta \setminus [M\rangle_\delta$. The inclusion

$$\mathcal{X} \subset [M\rangle_\delta \cup [U]_\delta$$

shows that $\mathcal{X}$ coincides with $[M\rangle_\delta$ or $[U]_\delta$.   □

Example 4.8. Let M be an m-dimensional singular subspace of Π. Then $[M\rangle_\delta$ is a subspace of $\mathfrak{G}_\delta(\Pi)$; subspaces of such type are called *parabolic* [Cooperstein, Kasikova and Shult (2005)]. In the case when $m < n - 4$, this subspace is non-singular. As in Example 4.6, we consider $[M\rangle_{m+1}$. This is a polar space of type D_{n-m-1}. The parabolic subspace $[M\rangle_\delta$ can be identified with one of the half-spin Grassmann spaces of this polar space.

Exercise 4.9. Show that the intersection of two distinct stars of $\mathfrak{G}_\delta(\Pi)$ is empty, or a single point or a line (the second possibility is not realized if $n = 4$); moreover, this intersection is a line if and only if the associated $(n - 4)$-dimensional singular subspaces are adjacent.

Exercise 4.10. Show that the intersection of two distinct special subspaces of $\mathfrak{G}_\delta(\Pi)$ is empty or a line, the second possibility is realized if and only if the associated elements of $\mathcal{G}_{-\delta}(\Pi)$ are adjacent.

If $[M\rangle_\delta$ and $[U]_\delta$ are a star and a special subspace satisfying $M \subset U$ then their intersection is a plane (see the proof of Proposition 4.20).

Exercise 4.11. Show that every plane of $\mathfrak{G}_\delta(\Pi)$ is contained in precisely one star and precisely one special subspace. *Hint:* take any triangle in a plane and consider the associated elements of $\mathcal{G}_{n-4}(\Pi)$ and $\mathcal{G}_{-\delta}(\Pi)$ (Lemma 4.11).

As above, we write $\mathcal{X}^\sim$ for the set consisting of all elements of $\mathcal{G}_\delta(\Pi)$ adjacent with all elements of $\mathcal{X} \subset \mathcal{G}_\delta(\Pi)$. We have the standard characterization of lines in terms of the adjacency relation.

Proposition 4.22. *For any adjacent* $S, U \in \mathcal{G}_\delta(\Pi)$ *the line joining* S *and* U *coincides with the set* $\{S, U\}^{\sim\sim}$.

Proof. The intersection of all maximal singular subspaces of $\mathfrak{G}_\delta(\Pi)$ (stars and special subspaces) containing both S and U coincides with the line joining S and U.   □

A few remarks concerning polar spaces of type D_4 finish the subsection.

Proposition 4.23. *If $n = 4$ then $\mathfrak{G}_\delta(\Pi)$ is a polar space of type D_4.*

Proof. The polar axioms (1), (3) and (4) hold and we verify (2). Let S be a point of $\mathfrak{G}_\delta(\Pi)$ which does not belong to a line $[L\rangle_\delta$, $L \in \mathcal{L}$. If S has a non-empty intersection with L then it is collinear with all points of $[L\rangle_\delta$. If L and S are disjoint then $S \cap L^\perp$ is a line of Π (Lemma 4.2) and

$$\langle L, S \cap L^\perp \rangle$$

is the unique point on the line $[L\rangle_\delta$ collinear with S. Therefore, $\mathfrak{G}_\delta(\Pi)$ is a polar space. Maximal singular subspaces of $\mathfrak{G}_\delta(\Pi)$ (stars and special subspaces) are 3-dimensional and every plane is contained in precisely one star and one special subspace (Exercise 4.11). The polar space is of type D_4. $\qquad\square$

Exercise 4.12. Suppose that $n = 4$. Show that the natural one-to-one correspondence between lines of $\mathfrak{G}_\delta(\Pi)$ and lines of Π is a collineation of $\mathfrak{G}_1(\mathfrak{G}_\delta(\Pi))$ to $\mathfrak{G}_1(\Pi)$. Consider the half-spin Grassmann spaces of the polar space $\mathfrak{G}_\delta(\Pi)$: one of them consists of stars and the other is formed by special subspaces; show that they are isomorphic to Π and $\mathfrak{G}_{-\delta}(\Pi)$, respectively.

4.6 Collineations

Throughout the section we suppose that $\Pi = (P, \mathcal{L})$ and $\Pi' = (P', \mathcal{L}')$ are polar spaces of same type X_n, $X \in \{C, D\}$ and $n \geq 3$; in the case when $X = D$, we will require that $n \geq 4$.

4.6.1 *Chow's theorem and its generalizations*

Every collineation of Π to Π' induces a collineation of $\mathfrak{G}_k(\Pi)$ to $\mathfrak{G}_k(\Pi')$ for each $k \in \{1, \ldots, n-1\}$. Moreover, if our polar spaces are of type D_n then we get a collineation of $\mathfrak{G}_\delta(\Pi)$ to $\mathfrak{G}_\gamma(\Pi')$ with $\delta, \gamma \in \{+, -\}$.

There is the following analogue of Theorem 3.2.

Theorem 4.7 ([Chow (1949)]). *Every isomorphism of $\Gamma_{n-1}(\Pi)$ to $\Gamma_{n-1}(\Pi')$ is the collineation of $\mathfrak{G}_{n-1}(\Pi)$ to $\mathfrak{G}_{n-1}(\Pi')$ induced by a collineation of Π to Π'. Moveover, if our polar spaces are of type D_n ($n \geq 5$) then every isomorphism of $\Gamma_\delta(\Pi)$ to $\Gamma_\gamma(\Pi')$, $\delta, \gamma \in \{+, -\}$, is the collineation of $\mathfrak{G}_\delta(\Pi)$ to $\mathfrak{G}_\gamma(\Pi')$ induced by a collineation of Π to Π'.*

Remark 4.9. In [Chow (1949)] this result was established only for the polar spaces associated with reflexive forms, but the method works in the general case.

Remark 4.10. Suppose that Π and Π' both are symplectic polar spaces and write V and V' for the associated $(2n)$-dimensional vector spaces. It was proved in [Huang (2000)] that every surjection of $\mathcal{G}_{n-1}(\Pi)$ to $\mathcal{G}_{n-1}(\Pi')$ sending adjacent elements to adjacent elements is a collineation of $\mathfrak{G}_{n-1}(\Pi)$ to $\mathfrak{G}_{n-1}(\Pi')$; in particular, every semicollineation of $\mathfrak{G}_{n-1}(\Pi)$ to $\mathfrak{G}_{n-1}(\Pi')$ is a collineation. The following more general version of this result was established in [Huang (2001)]: if a mapping

$$f : \mathcal{G}_{n-1}(\Pi) \to \mathcal{G}_{n-1}(\Pi')$$

sends adjacent elements to adjacent elements and for every $S \in \mathcal{G}_{n-1}(\Pi)$ there exists $U \in \mathcal{G}_{n-1}(\Pi)$ such that the distance between $f(S)$ and $f(U)$ is maximal then f can be extended to the embedding of $\mathfrak{G}_n(V)$ in $\mathfrak{G}_n(V')$ induced by a semilinear embedding of V in V'.

Theorem 4.8 ([Pankov, Prażmowski and Żynel (2006)]). *Let f be an isomorphism of $\Gamma_k(\Pi)$ to $\Gamma_k(\Pi')$ and $0 \le k \le n-2$. If*

$$n \ne 4 \quad or \quad k \ne 1$$

then f is the collineation of $\mathfrak{G}_k(\Pi)$ to $\mathfrak{G}_k(\Pi')$ induced by a collineation of Π to Π' (f is a collineation of Π to Π' if $k = 0$).

In Subsection 4.6.3 we establish that an isomorphism of $\Gamma_k(\Pi)$ to $\Gamma_k(\Pi')$ ($k \le n-3$) is induced by a collineation of Π to Π' if it transfers stars to stars and tops to tops. It follows from elementary properties of triangles that the latter condition holds for $n \ge 5$. Using Chow's idea, we show that every isomorphism of $\Gamma_{n-2}(\Pi)$ to $\Gamma_{n-2}(\Pi')$ induces an isomorphism of $\Gamma_{n-3}(\Pi)$ to $\Gamma_{n-3}(\Pi')$ which sends stars to stars and tops to tops (Subsection 4.6.4). This is a modification of the proof given in [Pankov, Prażmowski and Żynel (2006)]. Theorem 4.7 will be presented as a simple consequence of Theorem 4.8.

Now consider the remaining case $n = 4$ and $k = 1$. Suppose that our polar spaces are of type D_4. Then the half-spin Grassmann spaces are polar spaces of type D_4 and there is natural one-to-one correspondence between their lines and lines of the polar spaces (every line of $\mathfrak{G}_\delta(\Pi)$, $\delta \in \{+, -\}$, consists of all elements of $\mathcal{G}_\delta(\Pi)$ containing a certain line of Π). This means that a collineation f of Π to $\mathfrak{G}_\delta(\Pi')$ (if it exists) induces a bijection of $\mathcal{L}$ to

$\mathcal{L}'$; an easy verification shows that this is a collineation of $\mathfrak{G}_1(\Pi)$ to $\mathfrak{G}_1(\Pi')$. It will be shown later that such collineations map all tops to stars and some stars to tops. The collineation f also induces a collineation between one of the half-spin Grassmann spaces of Π and $\mathfrak{G}_{-\delta}(\Pi')$ (since $\mathfrak{G}_{-\delta}(\Pi')$ is one of the half-spin Grassmann spaces of the polar space $\mathfrak{G}_\delta(\Pi')$, see Exercise 4.12); this collineation sends stars to special subspaces and special subspaces to stars.

Theorem 4.9. *If $n = 4$ then every isomorphism of $\Gamma_1(\Pi)$ to $\Gamma_1(\Pi')$ is the collineation of $\mathfrak{G}_1(\Pi)$ to $\mathfrak{G}_1(\Pi')$ induced by a collineation of Π to Π' or a collineation of Π to one of the half-spin Grassmann spaces of Π'; the second possibility can be realized only in the case when the polar spaces are of type* D_4.

Theorem 4.10. *If Π and Π' are of type D_4 then every isomorphism of $\Gamma_\delta(\Pi)$ to $\Gamma_\gamma(\Pi')$, $\delta, \gamma \in \{+, -\}$, is the collineation of $\mathfrak{G}_\delta(\Pi)$ to $\mathfrak{G}_\gamma(\Pi')$ induced by a collineation of Π to Π' or a collineation of Π to $\mathfrak{G}_{-\gamma}(\Pi')$.*

4.6.2 *Weak adjacency on polar Grassmannians*

To prove Theorems 4.7 and 4.8 we will use elementary properties of so-called *weak adjacency* relation defined on $\mathcal{G}_k(\Pi)$, $0 < k < n - 1$. In Section 4.9 these properties also will be exploited to study apartments preserving mappings.

Two elements of the Grassmannian $\mathcal{G}_k(\Pi)$, $0 < k < n - 1$, are said to be *weakly adjacent* if their intersection belongs to $\mathcal{G}_{k-1}(\Pi)$. By this definition, any two adjacent elements of $\mathcal{G}_k(\Pi)$ are weakly adjacent; the converse fails. The *weak Grassmann graph* $\Gamma_k^w(\Pi)$ is the graph whose vertex set is $\mathcal{G}_k(\Pi)$ and whose edges are pairs of weakly adjacent elements. Since $\Gamma_k(\Pi)$ is a subgraph of $\Gamma_k^w(\Pi)$, the weak Grassmann graph is connected and every clique of $\Gamma_k(\Pi)$ is a clique in $\Gamma_k^w(\Pi)$.

Example 4.9. As in Example 3.1, we show that every top of $\mathcal{G}_k(\Pi)$ is a maximal clique of $\Gamma_k^w(\Pi)$.

Example 4.10. For every $M \in \mathcal{G}_{k-1}(\Pi)$ the parabolic subspace $[M\rangle_k$ will be called a *big star*. This is a clique of $\Gamma_k^w(\Pi)$ containing non-adjacent elements. If S_1 and S_2 are distinct non-adjacent elements of $[M\rangle_k$ and $S \in \mathcal{G}_k(\Pi)$ is weakly adjacent with both S_1, S_2 then S belongs to $[M\rangle_k$. Indeed, if S intersects S_1 and S_2 in distinct $(k - 1)$-dimensional subspaces then we take points

$$p_i \in (S \cap S_i) \setminus (S_1 \cap S_2), \quad i = 1, 2;$$

since $p_1 \perp p_2$ (these points belong to S) and

$$S_i = \langle S_1 \cap S_2, p_i \rangle,$$

we get $S_1 \perp S_2$ which contradicts the assumption that S_1 and S_2 are non-adjacent. Therefore, big stars are maximal cliques of $\Gamma_k^w(\Pi)$.

Proposition 4.24. *Every maximal clique of $\Gamma_k^w(\Pi)$ is a top or a big star.*

Proof. Let $\mathcal{X}$ be a maximal clique of $\Gamma_k^w(\Pi)$. If any two distinct elements of $\mathcal{X}$ are adjacent then it is a maximal clique of $\Gamma_k(\Pi)$; hence $\mathcal{X}$ is a top (stars are non-maximal cliques of $\Gamma_k^w(\Pi)$, since they are proper subsets of big stars). If S_1 and S_2 are distinct non-adjacent elements of $\mathcal{X}$ then every element of $\mathcal{X}$ contains $S_1 \cap S_2$ (see Example 4.10) and $\mathcal{X}$ is the big star corresponding to $S_1 \cap S_2$. □

4.6.3 Proof of Theorem 4.8 for $k < n - 2$

Let f be an isomorphism of $\Gamma_k(\Pi)$ to $\Gamma_k(\Pi')$ and $k < n - 2$. It follows directly from the characterization of lines in terms of the adjacency relation (Proposition 4.17) that f is a collineation of $\mathfrak{G}_k(\Pi)$ to $\mathfrak{G}_k(\Pi')$ and Theorem 4.8 is true for $k = 0$. If $k > 0$ then f and f^{-1} map maximal singular subspaces (stars and tops) to maximal singular subspaces.

Lemma 4.12. *If f is a collineation of $\mathfrak{G}_k(\Pi)$ to $\mathfrak{G}_k(\Pi')$, $k < n-2$, transferring tops to tops and stars to stars then it is induced by a collineation of Π to Π'.*

Proof. Two stars $[M, N]_k$ and $[M', N']_k$ are called *adjacent* if $M = M'$ and N is adjacent with N'. Two distinct stars $\mathcal{X}, \mathcal{X}'$ are non-adjacent if and only if their intersection is empty or there exist distinct stars $\mathcal{Y}, \mathcal{Y}'$ such that $\mathcal{X} \cap \mathcal{X}'$ is a proper subset of $\mathcal{Y} \cap \mathcal{Y}'$. In other words, pairs of adjacent stars can be characterized as pairs with maximal intersections. This means that f and f^{-1} maps adjacent stars to adjacent stars.

For every $M \in \mathcal{G}_{k-1}(\Pi)$

$$[M\rangle_k = \bigcup_{N \in [M\rangle_{n-1}} [M, N]_k.$$

Any two elements of $[M\rangle_{n-1}$ can be connected by a path of $\Gamma_{n-1}(\Pi)$ contained in $[M\rangle_{n-1}$ (the parabolic subspace $[M\rangle_{n-1}$ is isomorphic to the Grassmann space consisting of maximal singular subspaces of a certain polar space, Example 4.6). This implies the existence of a subspace

$$f_{k-1}(M) \in \mathcal{G}_{k-1}(\Pi')$$

such that the associated big star of $\mathcal{G}_k(\Pi')$ contains $f([M\rangle_k)$. It is clear that

$$f([M\rangle_k) = [f_{k-1}(M)\rangle_k$$

for every $M \in \mathcal{G}_{k-1}(\Pi)$ (we apply the same arguments to f^{-1}) and the mapping

$$f_{k-1} : \mathcal{G}_{k-1}(\Pi) \to \mathcal{G}_{k-1}(\Pi')$$

is bijective. Since for every $S \in \mathcal{G}_k(\Pi)$

$$M \in \langle S]_{k-1} \Leftrightarrow S \in [M\rangle_k \Leftrightarrow f(S) \in [f_{k-1}(M)\rangle_k \Leftrightarrow f_{k-1}(M) \in \langle f(S)]_{k-1},$$

we have

$$f_{k-1}(\langle S]_{k-1}) = \langle f(S)]_{k-1}.$$

Therefore, f_{k-1} is an isomorphism of $\Gamma_{k-1}(\Pi)$ to $\Gamma_{k-1}(\Pi')$ such that f_{k-1} and the inverse mapping send tops to tops (this is a collineation of Π to Π' if $k = 1$). If $k \geq 2$ then f_{k-1} maps stars to stars. As in the proof of Theorem 3.2, we get a sequence of collineations

$$f_i : \mathcal{G}_i(\Pi) \to \mathcal{G}_i(\Pi'), \quad i = k, \ldots, 0,$$

such that $f_k = f$ and establish that every f_i is induced by f_0. $\qquad\Box$

So, we need to show that f maps stars to stars and tops to tops, except the case when $n = 4$ and $k = 1$. Since the dimension of stars and tops is equal to

$$n - k - 1 \text{ and } k + 1$$

(respectively), this is true if $n \neq 2k + 2$.

Suppose that $n = 2k + 2$. Every top-triangle is contained in precisely one maximal singular subspace (a top). If $S_1, S_2, S_3 \in \mathcal{G}_k(\Pi)$ form a star-triangle then the singular subspace

$$\langle S_1, S_2, S_3 \rangle$$

is $(k + 2)$-dimensional; in the case when $k > 1$, this singular subspace is not maximal and the star-triangle is contained in more than one star. Therefore, if $k > 1$ then star-triangles go to star-triangles, top-triangles go to top-triangles and we get the claim.

4.6.4 *Proof of Theorems 4.7 and 4.8*

Let f be an isomorphism of $\Gamma_{n-2}(\Pi)$ to $\Gamma_{n-2}(\Pi')$. Maximal cliques of these graphs are tops and f induces a bijection

$$g : \mathcal{G}_{n-1}(\Pi) \to \mathcal{G}_{n-1}(\Pi').$$

This mapping transfers a line $[S\rangle_{n-1}$, $S \in \mathcal{G}_{n-2}(\Pi)$, to the line $[f(S)\rangle_{n-1}$. Hence g is a collineation of $\mathfrak{G}_{n-1}(\Pi)$ to $\mathfrak{G}_{n-1}(\Pi')$ and

$$\dim(g(M) \cap g(N)) = \dim(M \cap N)$$

for all $M, N \in \mathcal{G}_{n-1}(\Pi)$ (the latter equality follows from the distance formula given in Proposition 4.8).

Now we show that f is an isomorphism of $\Gamma^w_{n-2}(\Pi)$ to $\Gamma^w_{n-2}(\Pi')$.

Let S_1, S_2 be weakly adjacent elements of $\mathcal{G}_{n-2}(\Pi)$. Let also M'_1, M'_2 be maxima singular subspaces of Π' such that

$$M'_1 \cap M'_2 = f(S_1) \cap f(S_2)$$

and $f(S_i) \subset M'_i$ for $i = 1, 2$ (Corollary 4.2). If $f(S_1)$ and $f(S_2)$ are not weakly adjacent then

$$\dim(M'_1 \cap M'_2) < n - 3$$

and

$$\dim(g^{-1}(M'_1) \cap g^{-1}(M'_2)) < n - 3.$$

Since g is induced by f, we have

$$S_i \subset g^{-1}(M'_i), \quad i = 1, 2,$$

and the latter inequality contradicts the fact that S_1 and S_2 are weakly adjacent. Similarly, we establish that f^{-1} preserves the weak adjacency relation.

Since f and f^{-1} transfer tops to tops, big stars go to big stars in both directions. This means that f induces a bijection

$$h : \mathcal{G}_{n-3}(\Pi) \to \mathcal{G}_{n-3}(\Pi').$$

In the case when $n = 3$, this is a collineation of Π to Π' which induces f. If $n \geq 4$ then h is a collineation of $\mathfrak{G}_{n-3}(\Pi)$ to $\mathfrak{G}_{n-3}(\Pi')$ preserving the types of all maximal singular subspaces. By Lemma 4.12, h is induced by a collineation of Π to Π'; this collineation induces f.

Theorem 4.8 is proved and we use it to prove Theorem 4.7.

Since maximal cliques of $\Gamma_{n-1}(\Pi)$ and $\Gamma_{n-1}(\Pi')$ are lines of the associated Grassmann spaces, every isomorphism t between these graphs induces a bijection $\tilde{t}$ of $\mathcal{G}_{n-2}(\Pi)$ to $\mathcal{G}_{n-2}(\Pi')$. This bijection is an isomorphism of $\Gamma_{n-2}(\Pi)$ to $\Gamma_{n-2}(\Pi')$ (two distinct lines have a non-empty intersection if and only if the associated $(n-2)$-dimensional singular subspaces are adjacent). By Theorem 4.8, $\tilde{t}$ is induced by a collineation of Π to Π'; this collineation induces t.

Now suppose that our polar spaces are of type D_n and f is an isomorphism of $\Gamma_\delta(\Pi)$ to $\Gamma_\gamma(\Pi')$, $\delta, \gamma \in \{+, -\}$. Then f is a collineation of $\mathfrak{G}_\delta(\Pi)$ to $\mathfrak{G}_\gamma(\Pi')$ (by Proposition 4.22, lines of half-spin Grassmann spaces can be characterized in terms of the adjacency relation). Hence f maps lines to lines and induces a bijection of $\mathcal{G}_{n-3}(\Pi)$ to $\mathcal{G}_{n-3}(\Pi')$. Since two distinct $(n-3)$-dimensional singular subspaces of Π or Π' are adjacent if and only if the associated lines of the half-spin Grassmann space $\mathfrak{G}_\delta(\Pi)$ or $\mathfrak{G}_\gamma(\Pi')$ span a plane, this bijection is an isomorphism of $\Gamma_{n-3}(\Pi)$ to $\Gamma_{n-3}(\Pi')$. In the case when $n \geq 5$, we apply Theorem 4.8 and get the claim.

Similarly, we draw Theorem 4.10 from Theorem 4.9.

4.6.5 *Proof of Theorem 4.9*

Throughout the subsection we assume that $n = 4$.

Suppose that f is an isomorphism of $\Gamma_1(\Pi)$ to $\Gamma_1(\Pi')$. It was noted in Subsection 4.6.3 that f is a collineation of $\mathfrak{G}_1(\Pi)$ to $\mathfrak{G}_1(\Pi')$; moreover, f is induced by a collineation of Π to Π' if it preserves the types of maximal singular subspaces (Lemma 4.12). Now we suppose that the image of a certain star is a top; under this assumption, we establish that our polar spaces are of type D_4 and f is induced by a collineation of Π to one of the half-spin Grassmann spaces of Π'. Therefore, if the polar spaces are of type C_4 then f and f^{-1} both map stars to stars (hence tops go to tops) and f is induced by a collineation of Π to Π'.

0. Preliminaries. Our proof will be based on some trivial observations concerning pairs of triangles. In the present case ($n = 4, k = 1$), every triangle is contained in precisely one maximal singular subspace.

We say that triangles

$$\Delta = \{L_1, L_2, L_3\} \quad \text{and} \quad \Delta' = \{L_1', L_2', L_3'\}, \qquad L_i \neq L_j' \quad \forall\, i, j, \qquad (4.3)$$

in $\mathfrak{G}_1(\Pi)$ form a *regular pair* if L_i and L_j' are adjacent only in the case when $i \neq j$; in other words, every line from each of these triangles is adjacent with precisely two lines of the other.

Lemma 4.13. *If the triangles* (4.3) *form a regular pair then one of the following possibilities is realized:*

- *There exist a plane S and a point $p \notin S$, $p \perp S$, such that one of the triangles is a star-triangle in $[p, \langle p, S \rangle]_1$ and the other is a top-triangle in $\langle S \rangle_1$.*
- *There exist maximal singular subspaces U, U' intersecting precisely in a point p and such that Δ and Δ' are star-triangles in $[p, U]_1$ and $[p, U']_1$, respectively.*

Proof.   Direct verification. $\square$

We say that triangles

$$\Delta = \{L, L_1, L_2\} \ \text{ and } \ \Delta' = \{L, L_1', L_2'\}, \qquad L_i \neq L_j' \ \ \forall \, i, j, \qquad (4.4)$$

in $\mathfrak{G}_1(\Pi)$ form a *regular pair* if L_i and L_j' are adjacent only in the case when $i = j$.

Lemma 4.14. *If the triangles (4.4) form a regular pair then one of the following possibilities is realized:*

- *There exist two planes S, S' intersecting precisely in L and such that $S \perp S'$ and Δ, Δ' are top-triangles in $\langle S \rangle_1$ and $\langle S' \rangle_1$, respectively.*
- *There exist two maximal singular subspaces U, U' intersecting precisely in L and a point $p \in L$ such that Δ, Δ' are star-triangles in $[p, U]_1$ and $[p, U']_1$, respectively.*

Proof.   Direct verification. $\square$

1. Let f be an injection of $\mathcal{G}_1(\Pi)$ to $\mathcal{G}_1(\Pi')$ preserving the adjacency relation: two elements of $\mathcal{G}_1(\Pi)$ are adjacent if and only if their images are adjacent. Then f transfers maximal singular subspaces of $\mathfrak{G}_1(\Pi)$ (stars and tops) to subsets of maximal singular subspaces of $\mathfrak{G}_1(\Pi')$. Every maximal singular subspace of $\mathfrak{G}_1(\Pi')$ contains at most one image of a maximal singular subspace of $\mathfrak{G}_1(\Pi)$ (otherwise there exist non-adjacent elements of $\mathcal{G}_1(\Pi)$ whose images are adjacent). We will assume that f satisfies the following condition:

(A) triangles go to triangles.

It is clear that two triangles in $\mathfrak{G}_1(\Pi)$ form a regular pair if and only if their images form a regular pair in $\mathfrak{G}_1(\Pi')$. The condition (A) guarantees that the image of every maximal singular subspace of $\mathfrak{G}_1(\Pi)$ is contained in precisely one maximal singular subspace of $\mathfrak{G}_1(\Pi')$ (since triangles are bases of maximal singular subspaces).

Suppose that the image of a certain star $[p, U]_1$, $U \in \mathcal{G}_3(\Pi), p \in U$, is contained in a top $\langle S' \rangle_1$, $S' \in \mathcal{G}_2(\Pi')$. Our first step is the following.

Lemma 4.15. *There exists a maximal singular subspace U' of Π' containing S' and satisfying the following conditions:*

- *for every point $q \in U$ there is a plane $S(q) \subset U'$ such that*

$$f([q, U]_1) \subset \langle S(q) \rangle_1,$$

- *for every plane $S \subset U$ there is a point $q(S) \in U'$ such that*

$$f(\langle S \rangle_1) \subset [q(S), U']_1.$$

Proof. Let S be a plane in U which does not contain the point p. Consider any regular pair of triangles

$$\Delta_1 \subset [p, U]_1 \quad \text{and} \quad \Delta_2 \subset \langle S \rangle_1.$$

Their images also form a regular pair in $\mathfrak{G}_1(\Pi')$. By our hypothesis, $f(\Delta_1)$ is a top-triangle contained in $\langle S' \rangle_1$. Then Lemma 4.13 guarantees that $f(\Delta_2)$ is a star-triangle; moreover, there exist a maximal singular subspace U' of Π' and a point $q(S) \in U'$ such that

$$q(S) \notin S' \subset U'$$

and $f(\Delta_2)$ is contained in the star $[q(S), U']_1$. The latter means that $f(\langle S \rangle_1)$ is a subset of $[q(S), U']_1$.

Similarly, for another plane $T \subset U$ which does not contain p, the image of the top $\langle T \rangle_1$ is contained in a certain star $[q(T), U'']_1$ and

$$q(T) \notin S' \subset U''.$$

If U' and U'' are distinct then their intersection is S'. Since the points $q(S)$ and $q(T)$ do not belong to S',

$$[q(S), U']_1 \cap [q(T), U'']_1 = \emptyset.$$

On the other hand,

$$f(\langle S \rangle_1) \subset [q(S), U']_1 \quad \text{and} \quad f(\langle T \rangle_1) \subset [q(T), U'']_1;$$

the tops $\langle S \rangle_1$ and $\langle T \rangle_1$ have a non-empty intersection (since $S \cap T$ is a line), a contradiction. Therefore, U'' coincides with U'.

Now, let $q \in U \setminus \{p\}$. We choose a plane $S \subset U$ which does not contain the points q and p. Then $f(\langle S \rangle_1)$ is contained in the star $[q(S), U']_1$ (it was established above). We consider any regular pair of triangles

$$\Delta_1 \subset [q, U]_1 \quad \text{and} \quad \Delta_2 \subset \langle S \rangle_1.$$

Their images form a regular pair and $f(\Delta_2)$ is a star-triangle in $[q(S), U']_1$. If $f(\Delta_1)$ is a star-triangle then, by Lemma 4.13, f transfers the star $[q, U]_1$ to a subset of a certain star $[q(S), U'']_1$ such that

$$U' \cap U'' = q(S).$$

Since $S' \subset U'$, the latter equality guarantees that

$$[q(S), U'']_1 \cap \langle S' \rangle_1 = \emptyset.$$

However, $[p, U]_1$ and $[q, U]_1$ have a non-empty intersection (the line qp) and their images are contained in $\langle S' \rangle_1$ and $[q(S), U'']_1$, respectively. Thus $f(\Delta_1)$ is a top-triangle. This means that $f([q, U]_1)$ is contained in a certain top $\langle S(q) \rangle_1$ with $S(q) \subset U'$.

Consider a plane $S \subset U$ containing the point p and any point $q \in U \setminus S$. Then $f([q, U]_1)$ is contained in the top $\langle S(q) \rangle_1$. Using Lemma 4.13, we establish that $f(\langle S \rangle_1)$ is a subset of a certain star $[q(S), U'']_1$ such that

$$q(S) \notin S(q) \subset U''.$$

We take any plane $T \subset U$ which does not contain the points p and q. The image of the top $\langle T \rangle_1$ is contained in the star $[q(T), U']_1$ and $q(T) \notin S(q)$. As above, we show that U'' coincides with U'. $\qquad\square$

Since distinct maximal singular subspaces go to subsets of distinct maximal singular subspaces, the mappings

$$q \to S(q) \quad\text{and}\quad S \to q(S)$$

(we define $S(p) := S'$) are injective.

In the case when f is a collineation of $\mathfrak{G}_1(\Pi)$ to $\mathfrak{G}_1(\Pi')$, the inclusions in Lemma 4.15 must be replaced by the equalities. Moreover, the inverse mapping f^{-1} sends every star

$$[q(S), U']_1, \ S \in \langle U \rangle_2,$$

to the top $\langle S \rangle_1$. We apply the arguments from the proof of Lemma 4.15 to f^{-1} and establish that $q \to S(q)$ is a one-to-one correspondence between points of U and planes of U'; similarly, $S \to q(S)$ is a one-to-one correspondence between planes of U and points of U'.

2. Suppose that f is a collineation of $\mathfrak{G}_1(\Pi)$ to $\mathfrak{G}_1(\Pi')$ and U is as in the previous step. Let Q be a maximal singular subspace of Π intersecting U in a certain plane S. The image of the top $\langle S \rangle_1$ is the star $[q(S), U']_1$.

Let $q \in Q \setminus S$. We take any regular pair of triangles

$$\Delta_1 \subset \langle S \rangle_1 \quad\text{and}\quad \Delta_2 \subset [q, Q]_1.$$

Then $f(\Delta_1)$ is a star-triangle in $[q(S), U']_1$. By Lemma 4.13, there are the following two possibilities for the triangle $f(\Delta_2)$:

- $f(\Delta_2)$ is a top-triangle then $f([q,Q]_1)$ coincides with a certain top $\langle T'\rangle_1$ with $T' \subset U'$,
- $f(\Delta_2)$ is a star-triangle then $f([q,Q]_1)$ is a star $[q(S), Q']_1$ such that Q' intersects U' precisely in the point $q(S)$.

The first possibility is not realized, since for any plane $T' \subset U'$ the top $\langle T'\rangle_1$ is the image of a certain star $[t, U]_1$.

Now, suppose that $q \in S$. If the image of the star $[q,Q]_1$ is a top then, by the first step of our proof, the same holds for all points of Q which is impossible. Therefore, $f([q,Q]_1)$ is a star for every point $q \in Q$.

Let M be a third maximal singular subspace containing S. We take points $q \in Q \setminus S$ and $t \in M \setminus S$. Then every element of $[q,Q]_1$ is non-adjacent with every element of $[t, M]_1$. However,

$$f([q,Q]_1) = [q(S), Q']_1$$

contains elements adjacent with some elements of

$$f([t, M]_1) = [q(S), M']_1.$$

This means that there are no 3 distinct maximal singular subspaces containing S and *our polar spaces are of type* D_4.

3. Suppose that the polar spaces are of type D_4 and, as in the first step, f is an injection of $\mathcal{G}_1(\Pi)$ to $\mathcal{G}_1(\Pi')$ preserving the adjacency relation and satisfying the condition (A). We also assume that the maximal singular subspaces U and U' belong to $\mathcal{G}_\delta(\Pi)$ and $\mathcal{G}_\gamma(\Pi')$, $\delta, \gamma \in \{+, -\}$, respectively.

Let Q be an element of $\mathcal{G}_\delta(\Pi)$ adjacent with U. The intersection of U and Q is a certain line L. We take a point $q \in L$ and consider any regular pair of star-triangles

$$\Delta_1 = \{L, L_1, L_2\} \subset [q, U]_1 \text{ and } \Delta_2 = \{L, L_1', L_2'\} \subset [q, Q]_1.$$

The image of the star $[q, U]_1$ is contained in the top $\langle S(q)\rangle_1$ and $f(\Delta_1)$ is a top-triangle. By Lemma 4.14, $f(\Delta_2)$ also is a top-triangle which means that $f([q,Q]_1)$ is a subset of a certain top.

Lemma 4.15 implies the existence of a maximal singular subspace Q' of Π' satisfying the following conditions: for every point $q \in Q$ there is a plane $T(q) \subset Q'$ such that

$$f([q,Q]_1) \subset \langle T(q)\rangle_1,$$

and for every plane $S \subset Q$ there is a point $t(S) \in Q'$ such that

$$f(\langle S]_1) \subset [t(S), Q']_1.$$

Now we choose two planes $S \subset U$ and $M \subset Q$ intersecting in L and satisfying $S \perp M$. Consider any regular pair of top-triangles

$$\Delta_1 = \{L, L_1, L_2\} \subset \langle S \rangle_1 \text{ and } \Delta_2 = \{L, L_1', L_2'\} \subset \langle M \rangle_1.$$

Then

$$f(\Delta_1) \subset [q(S), U']_1 \text{ and } f(\Delta_2) \subset [t(M), Q']_1.$$

Since these star-triangles form a regular pair, Lemma 4.14 implies that $q(S) = t(M)$ and $U' \cap Q' = f(L)$. In particular, Q' is an element of $\mathcal{G}_\gamma(\Pi')$ adjacent with U'.

The half-spin Grassmann spaces are connected and the same holds for every $Q \in \mathcal{G}_\delta(\Pi)$. Thus we get a mapping

$$g : \mathcal{G}_\delta(\Pi) \to \mathcal{G}_\gamma(\Pi')$$

satisfying the following conditions: if $M \in \mathcal{G}_\delta(\Pi)$ then for any point $q \in M$ there is a plane $S_M(q) \subset g(M)$ such that

$$f([q, M]_1) \subset \langle S_M(q) \rangle_1,$$

and for any plane $S \subset M$ there is a point $q_M(S) \in g(M)$ such that

$$f(\langle S \rangle_1) \subset [q_M(S), g(M)]_1.$$

Remark 4.11. The mapping g sends adjacent elements to adjacent elements; but it does not need to be injective.

In the case when f is a collineation of $\mathfrak{G}_1(\Pi)$ to $\mathfrak{G}_1(\Pi')$, we apply the same arguments to f^{-1} and get the following:

- the latter two inclusions must be replaced by the equalities;
- $q \to S_M(q)$ is a one-to-one correspondence between points of M and planes of $g(M)$; similarly, $S \to q_M(S)$ is a one-to-one correspondence between planes of M and points of $g(M)$;
- the mapping g is bijective.

Since every plane of Π is contained in a certain element of $\mathcal{G}_\delta(\Pi)$, the images of all tops are subsets of stars (in the general case). If f is a collineation of $\mathfrak{G}_1(\Pi)$ to $\mathfrak{G}_1(\Pi')$ then f and f^{-1} map tops to stars; the maximal singular subspaces associated with these stars are elements of $\mathcal{G}_\delta(\Pi)$ and $\mathcal{G}_\gamma(\Pi')$, respectively.

4. As in the previous step, we assume that the polar spaces are of type D_4 and f is an injection of $\mathcal{G}_1(\Pi)$ to $\mathcal{G}_1(\Pi')$ preserving the adjacency relation and satisfying the condition (A).

We take any point $q \in P$ and consider the associated star $[q\rangle_\delta$ in the half-spin Grassmann space $\mathfrak{G}_\delta(\Pi)$. For every $M \in [q\rangle_\delta$ the image of the star $[q, M]_1$ is contained in the top $\langle S_M(q)]_1$. Let M and N be distinct elements of $[q\rangle_\delta$. The stars $[q, M]_1$ and $[q, N]_1$ have a non-empty intersection (the line $M \cap N$). This means that the planes $S_M(q)$ and $S_N(q)$ are weakly adjacent. Thus the set

$$\{S_M(q)\}_{M \in [q\rangle_\delta} \tag{4.5}$$

is a clique of the weak Grassmann graph $\Gamma_2^w(\Pi')$. Every element of $[q, M]_1$ is adjacent with a certain element of $[q, N]_1$ distinct from the line $M \cap N$. Hence every element of $\langle S_M(q)]_1$ is adjacent with an element of $\langle S_N(q)]_1$ distinct from $f(M \cap N)$. The latter guarantees that

$$S_M(q) \perp S_N(q)$$

and these planes are adjacent. Therefore, all elements of (4.5) are contained in a certain maximal singular subspace U_q (the set (4.5) contains more than one element and there is only one maximal singular subspace satisfying this condition). Since each

$$g(M) \in \mathcal{G}_\gamma(\Pi'), \ M \in [q\rangle_\delta,$$

intersects U_q in the plane $S_M(q)$, we have $U_q \in \mathcal{G}_{-\gamma}(\Pi')$.

Denote by h the mapping of P to $\mathcal{G}_{-\gamma}(\Pi')$ which sends every point $q \in P$ to the maximal singular subspace U_q.

5. Now, assume that f is a collineation of $\mathfrak{G}_1(\Pi)$ to $\mathfrak{G}_1(\Pi')$.

Let $\mathcal{X}$ be a maximal clique of $\Gamma_2^w(\Pi')$ containing (4.5). For every element of $\mathcal{X}$ the mapping f^{-1} transfers the associated top to a star

$$[t, T]_1, \ T \in \mathcal{G}_\delta(\Pi).$$

It has a non-empty intersection with every star

$$[q, M]_1, \ M \in [q\rangle_\delta,$$

which implies that $t = q$. Hence T belongs to $[q\rangle_\delta$ and $\mathcal{X}$ coincides with (4.5). This maximal clique is the top $\langle U_q]_2$.

The latter guarantees that the mapping h is injective. Show that h is bijective.

Let $T' \in \mathcal{G}_{-\gamma}(\Pi')$ and S_1', S_2' be distinct planes contained in T'. The intersection of these planes is a certain line L'. The mapping f^{-1} transfers the tops $\langle S_i']_1$ to stars $[q_i, Q_i]_1$, $i = 1, 2$; both Q_i are elements of $\mathcal{G}_\delta(\Pi)$. For any regular pair of top-triangles

$$\Delta_1 = \{L', L_1, L_2\} \subset \langle S_1']_1 \ \text{and} \ \Delta_2 = \{L', L_1', L_2'\} \subset \langle S_2']_1$$

the star-triangles

$$f^{-1}(\Delta_1) \subset [q_1, Q_1]_1 \text{ and } f^{-1}(\Delta_2) \subset [q_2, Q_2]_1$$

form a regular pair. By Lemma 4.14, we have $q_1 = q_2$. Therefore, $T' = U_q$ for a certain point $q \in P$ and we get the claim.

Let L be a line of Π. For every point $q \in L$ there exists a star

$$[q, Q]_1, \ Q \in \mathcal{G}_\delta(\Pi),$$

containing L. The image of this star is a top $\langle M']_1$ with $M' \subset U_q$ and we obtain that $f(L) \subset U_q$. Conversely, suppose that the line $f(L)$ is contained in U_q. Then $f(L)$ is an element of a certain top $\langle M']_1$, $M' \subset U_q$. The mapping f^{-1} sends this top to a star $[q, Q]_1$; since L belongs to this star, we have $q \in L$.

Thus h transfers every line $L \in \mathcal{L}$ to the line of $\mathfrak{G}_{-\gamma}(\Pi')$ associated with the line $f(L)$. This means that h is a collineation of Π to $\mathfrak{G}_{-\gamma}(\Pi')$ and f is induced by this collineation.

Remark 4.12. In the case when f is not a collineation, the mapping h transfers every line $L \in \mathcal{L}$ to a subset of the line of $\mathfrak{G}_{-\gamma}(\Pi')$ associated with $f(L)$.

4.6.6 *Remarks*

In this subsection we establish some results closely related with Theorems 4.8 and 4.9.

Proposition 4.25. *Let $k, m \in \{0, \ldots, n-1\}$ be distinct numbers. Suppose that*

$$g_k : \mathcal{G}_k(\Pi) \to \mathcal{G}_k(\Pi') \text{ and } g_m : \mathcal{G}_m(\Pi) \to \mathcal{G}_m(\Pi')$$

are bijections such that $S \in \mathcal{G}_k(\Pi)$ and $U \in \mathcal{G}_m(\Pi)$ are incident if and only if $g_k(S)$ and $g_m(U)$ are incident. Then g_k and g_m are induced by the same collineation of Π to Π'.

Proof. Similar to Proposition 3.4. □

As a consequence, we get the following.

Theorem 4.11 ([Pankov, Prażmowski and Żynel (2006)]). *Suppose that a bijection*

$$f : \mathcal{G}_k(\Pi) \to \mathcal{G}_k(\Pi'), \quad k < n-1,$$

preserves the relation $\perp$:

$$S \perp U \iff f(S) \perp f(U).$$

Then it is the collineation of $\mathfrak{G}_k(\Pi)$ *to* $\mathfrak{G}_k(\Pi')$ *induced by a collineation of* Π *to* Π'.

Proof. Every subspaces $\langle S]_k$, $S \in \mathcal{G}_{n-1}(\Pi)$, can be characterized as a maximal subset $\mathcal{X} \subset \mathcal{G}_k(\Pi)$ satisfying the following condition:

$$S \perp U \quad \forall\, S, U \in \mathcal{X}.$$

This means that f induces a bijection of $\mathcal{G}_{n-1}(\Pi)$ to $\mathcal{G}_{n-1}(\Pi')$ and we get a pair of bijections satisfying the condition of Proposition 4.25. $\square$

We will investigate isomorphisms between the weak Grassmann graphs and need the following lemma.

Lemma 4.16. *Two distinct elements* $S, U \in \mathcal{G}_k(\Pi)$, $1 \le k \le n - 2$, *are adjacent if and only if they belong to the intersection of two distinct maximal cliques of* $\Gamma_k^w(\Pi)$.

Proof. If S and U are adjacent then the big star $[S \cap U\rangle_k$ and the top $\langle\langle S, U\rangle]_k$ are as required. Conversely, suppose that there are two distinct maximal cliques of $\Gamma_k^w(\Pi)$ containing both S and U. Since the intersection of two distinct maximal cliques of the same type (two tops or two big stars) is empty or a single point, one of these cliques is a big star and the other is a top. This implies that S and U are adjacent. $\square$

Theorem 4.12 ([Pankov, Prażmowski and Żynel (2006)]). *Let* $1 \le k \le n - 2$. *Then every isomorphism of* $\Gamma_k^w(\Pi)$ *to* $\Gamma_k^w(\Pi')$ *is induced by a collineation of* Π *to* Π'.

Proof. Let f be an isomorphism of $\Gamma_k^w(\Pi)$ to $\Gamma_k^w(\Pi')$. Then f and f^{-1} transfer maximal cliques (tops and big stars) to maximal cliques. By Lemma 4.16, f is an isomorphism of $\Gamma_k(\Pi)$ to $\Gamma_k(\Pi')$. Therefore, it is induced by a collineation of Π to Π' if $n \ne 4$ or $k \ne 1$. Consider the case when $n = 4$ and $k = 1$. The mappings f and f^{-1} send tops to tops and big stars to big stars (since any two distinct elements of a top are adjacent and a big star contains non-adjacent elements). Hence f induces a bijection of P to P'. This is the required collineation of Π to Π'. $\square$

Remark 4.13. The collineation of $\mathfrak{G}_1(\Pi)$ to $\mathfrak{G}_1(\Pi')$ considered in the previous subsection (the case when Π and Π' are of type D_4) does not preserve the relation $\perp$ and it is not an isomorphism between the weak Grassmann graphs. For every point $q \in P$ it transfers the big stars $[q\rangle_1$ to $\langle U_q]_1$, where U_q is a maximal singular subspace of Π'.

4.7 Opposite relation

As for Grassmannians of finite-dimensional vector spaces, two elements of a polar or half-spin Grassmannian are said to be *opposite* if the distance between them is maximal (is equal to the diameter of the associated Grassmann graph).

4.7.1 *Opposite relation on polar Grassmannians*

Let Π be a polar space of rank n. Two points of Π (elements of $\mathcal{G}_0(\Pi)$) are opposite if and only if they are non-collinear. Two maximal singular subspaces are opposite if and only if they are disjoint.

The direct analogue of Theorem 3.6 does not hold in the general case; in other words, the following two conditions are not equivalent:

(1) $S_1, S_2 \in \mathcal{G}_k(\Pi)$ are adjacent,
(2) there exists $S \in \mathcal{G}_k(\Pi) \setminus \{S_1, S_2\}$ such that every $U \in \mathcal{G}_k(\Pi)$ opposite to S is opposite to at least one of S_i.

Example 4.11 ([Kwiatkowski and Pankov (2009)]). We consider a $(2n)$-dimensional vector space V over a field R of characteristic 2 and a non-degenerate alternating form Ω defined on V. Let

$$p_1, \ldots, p_n, q_1, \ldots, q_n$$

be a frame of the polar space Π_Ω such that $p_i \not\perp q_i$ for each i. There exist vectors $x_1, \ldots, x_n, y_1, \ldots, y_n \in V$ satisfying

$$p_i = \langle x_i \rangle, \ q_i = \langle y_i \rangle \ \text{ and } \ \Omega(x_i, y_i) = \Omega(y_i, x_i) = 1.$$

The maximal singular subspaces of Π_Ω corresponding to the maximal totally isotropic subspaces

$$\langle x_1, x_2, x_3, \ldots, x_n \rangle, \ \langle y_1, y_2, x_3, \ldots, x_n \rangle,$$

and

$$\langle x_1 + y_2, x_2 + y_1, x_3, \ldots, x_n \rangle$$

will be denoted by S_1, S_2, and S (respectively). Their intersection is the $(n-3)$-dimensional singular subspace N associated with the totally isotropic subspace $\langle x_3, \ldots, x_n \rangle$. We assert that for any collinear points

$$p' \in S_1 \setminus N \quad \text{and} \quad q' \in S_2 \setminus N$$

the line $p'q'$ intersects S. Indeed,

$$p' = \langle ax_1 + bx_2 + z' \rangle \quad \text{and} \quad q' = \langle ay_2 + by_1 + z'' \rangle$$

for some $a, b \in R$ and $z', z'' \in N$; every point of the line $p'q'$ is of type

$$\langle s(ax_1 + bx_2) + t(ay_2 + by_1) + z \rangle$$

with $z \in N$ and $s, t \in R$; this point belongs to S if $s = t = 1$. Thus if $U \in \mathcal{G}_{n-1}(\Pi_\Omega)$ intersects both S_1 and S_2 then it intersects S. This means that every element of $\mathcal{G}_{n-1}(\Pi_\Omega)$ opposite to S is opposite to at least one of S_i. However, S_1 and S_2 are not adjacent. In the case when the characteristic is not equal to 2, this construction is impossible.

Remark 4.14 (M. Kwiatkowski). Let p and q be distinct points of Π. We say that the Veldkamp line (the line in the associated Veldkamp space, Subsection 4.3.4) joining $p^\perp$ and $q^\perp$ is *thick* if it contains $t^\perp$, $t \neq p, q$; otherwise, it is said to be *thin*. Clearly, the line is thick if $p \perp q$; and it is thin if $p \not\perp q$ and Π is of type D_n. It is not difficult to show that all Veldkamp lines are thick if Π is the polar space defined by an alternating or Hermitian form. The points p and q satisfy (2) if and only if the Veldkamp line joining $p^\perp$ and $q^\perp$ is thick: we take any point t such that $t^\perp$ belongs to the Veldkamp line, then every point non-collinear with t is non-collinear with p or q. Thus the conditions (1) and (2) are not equivalent if $k = 0$ and Π is, for example, the polar space associated with an alternating or Hermitian form.

4.7.2 *Opposite relation on half-spin Grassmannians*

The direct analogue of Theorem 3.6 holds for half-spin Grassmannians.

Theorem 4.13 ([Kwiatkowski and Pankov (2009)]). *If Π is a polar space of type D_n, $n \geq 4$, then for any distinct $S_1, S_2 \in \mathcal{G}_\delta(\Pi)$, $\delta \in \{+, -\}$, the following conditions are equivalent:*

(1) S_1 *and* S_2 *are adjacent,*

(2) *there exists* $S \in \mathcal{G}_\delta(\Pi) \setminus \{S_1, S_2\}$ *such that every* $U \in \mathcal{G}_\delta(\Pi)$ *opposite to S is opposite to at least one of S_i.*

Corollary 4.3. *If* Π *and* Π' *are polar spaces of type* D_n, $n \geq 4$, *then every bijection of* $\mathcal{G}_\delta(\Pi)$ *to* $\mathcal{G}_\gamma(\Pi')$, $\delta, \gamma \in \{+, -\}$, *preserving the relation to be opposite* $(S, U \in \mathcal{G}_\delta(\Pi)$ *are opposite if and only if their images are opposite) is a collineation of* $\mathfrak{G}_\delta(\Pi)$ *to* $\mathfrak{G}_\gamma(\Pi')$.

Our proof of Theorem 4.13 is a modification of the proof of Theorem 3.6 given in Subsection 3.2.4; a verification of the conditions proposed in [Huang and Havlicek (2008)] (see Remark 3.6) is not simpler.

Proof. We will distinguish the following two cases:

- n is even then the dimension of the intersection of two elements of $\mathcal{G}_\delta(\Pi)$ is odd and $S, U \in \mathcal{G}_\delta(\Pi)$ are opposite if and only if $S \cap U = \emptyset$;
- n is odd then the dimension of the intersection of two elements of $\mathcal{G}_\delta(\Pi)$ is even and $S, U \in \mathcal{G}_\delta(\Pi)$ are opposite if and only if $S \cap U$ is a single point.

(1) $\Longrightarrow$ (2). Suppose that S_1, S_2 are adjacent and show that every $S \neq S_1, S_2$ belonging to the line of $\mathfrak{G}_\delta(\Pi)$ joining S_1 and S_2 is as required. Suppose that $U \in \mathcal{G}_\delta(\Pi)$ is opposite to S, but it is non-opposite to each S_i.

If n is even then U intersects S_1 and S_2 in subspaces whose dimensions are not less than 1. We take lines L_i contained in $U \cap S_i$, $i = 1, 2$. Since $S_1 \cap S_2 \subset S$ and S is opposite to U, these lines do not intersect $S_1 \cap S_2$. Hence S_i is spanned by $S_1 \cap S_2$ and L_i. The latter means that $L_1 \not\perp L_2$ which contradicts the fact that these lines are contained in U.

Consider the case when n is odd. According our assumption, the dimensions of $U \cap S_1$ and $U \cap S_2$ are not less than 2. Let P_i be planes contained in $U \cap S_i$, $i = 1, 2$. Each of these planes has a non-empty intersections with $S_1 \cap S_2$ (because $S_1 \cap S_2$ is $(n-3)$-dimensional). Since U is opposite to S and $S_1 \cap S_2 \subset S$, these intersections both are 0-dimensional (points). This implies the existence of lines $L_i \subset P_i$ $(i = 1, 2)$ which do not intersect $S_1 \cap S_2$. As above, we get $L_1 \not\perp L_2$.

(2) $\Longrightarrow$ (1). Our first step is the following statement: for every distinct collinear points $p_i \in S_i$, $i = 1, 2$, the line $p_1 p_2$ intersects S.

Suppose that n is even. If $p_1 p_2$ does not intersect S then we take any frame of Π whose subsets span S and the line $p_1 p_2$. Let U be the maximal singular subspace spanned by points of the frame and disjoint from S. This is an element of $\mathcal{G}_\delta(\Pi)$ opposite to S. By our hypothesis, U does not intersect at least one of S_i which contradicts the inclusion $p_1 p_2 \subset U$.

Now consider the case when n is odd. The intersection of S_1 and S_2 is non-empty. We assert that there exists a plane containing $p_1 p_2$ and intersecting each S_i in a line. If $p_1 p_2$ does not intersect $S_1 \cap S_2$ or the subspace $S_1 \cap S_2$ contains more than one point (in the second case the dimension of $S_1 \cap S_2$ is not less than 2) then we take any point $t \in S_1 \cap S_2$ which does not belong to $p_1 p_2$; the plane spanned by $p_1 p_2$ and t is as required. If $S_1 \cap S_2$ is a single point belonging to the line $p_1 p_2$ then this point coincides with p_1 or p_2. Suppose that $S_1 \cap S_2 = \{p_1\}$. Then any point $t \in S_1 \setminus S_2$ collinear with p_2 gives the claim (such point exists, since $S_1 \cap p_2^{\perp}$ is $(n-2)$-dimensional and the dimension of $S_1 \cap S_2$ is equal to 0). The case when $S_1 \cap S_2 = \{p_2\}$ is similar.

Let P be a plane containing $p_1 p_2$ and intersecting each S_i in a line. If $\dim(P \cap S) \leq 0$ then we take a frame whose subsets span P and S; using this frame, we construct $U \in \mathcal{G}_\delta(\Pi)$ opposite to S and containing P. Then U is opposite to at least one of S_i. Since $P \subset U$, the latter contradicts the fact that P intersects each S_i in a line. Therefore, $P \cap S$ contains a line and this line intersects $p_1 p_2$ (since P is a plane).

Our second step is to prove the equality

$$\dim(S \cap S_i) = n - 3, \quad i = 1, 2.$$

Let us take a point $p \in S_2 \setminus S$. Then $S_1 \cap p^{\perp}$ is a hyperplane of S_1 or $p \in S_1 \cap S_2$ and $S_1 \cap p^{\perp}$ coincides with S_1. In the first case, we set $H := S_1 \cap p^{\perp}$; in the second case, we take any hyperplane $H \subset S_1$. Let u, v be distinct points on a line $L \subset H$. The lines up and vp intersect S in points u' and v', respectively. Since $p \notin S$, we have $p \neq u', v'$ and the points u', v' are distinct. The lines L and $u'v'$ both are contained in the plane $\langle L, p \rangle$; thus they have a non-empty intersection. The inclusion $u'v' \subset S$ guarantees that L intersects S. So, every line of H has a non-empty intersection with S. The subspace H is $(n-2)$-dimensional and the dimension of $S \cap H$ is not less than $n - 3$. Since $S \cap H \subset S \cap S_1$, the subspace $S \cap S_1$ is $(n-3)$-dimensional. Similarly, we show that $S \cap S_2$ is $(n-3)$-dimensional.

Now we establish the equality

$$\dim(S_1 \cap S_2) = n - 3$$

which completes our proof. Define

$$U := (S \cap S_1) \cap (S \cap S_2).$$

Since $S \cap S_1$ and $S \cap S_2$ are $(n-3)$-dimensional subspaces of S, there are the following three possibilities:

- $S \cap S_1 = S \cap S_2$ and U is $(n-3)$-dimensional,

- $\dim U = n - 4$,
- $\dim U = n - 5$ and S is spanned by $S \cap S_1$ and $S \cap S_2$.

By $U \subset S_1 \cap S_2$, the dimension of $S_1 \cap S_2$ is equal to $n - 3$ in the first and second cases. Moreover, if U is an $(n - 5)$-dimensional subspace distinct from $S_1 \cap S_2$ then $S_1 \cap S_2$ is $(n - 3)$-dimensional.

Suppose that $U = S_1 \cap S_2$ is $(n - 5)$-dimensional. We take any line $L \subset S_1 \setminus S$ and consider the subspace $L^\perp \cap S_2$.

If every point of this subspace belongs to S then $L^\perp \cap S_2$ is contained in $S \cap S_2$. The subspace $S \cap S_2$ is $(n - 3)$-dimensional and the dimension of $L^\perp \cap S_2$ is not less than $n - 3$; thus

$$S \cap S_2 = L^\perp \cap S_2.$$

Since S is spanned by $S \cap S_1$ and $S \cap S_2$, the later equality implies that $L \perp S$ which is impossible (S is a maximal singular subspace and $L \not\subset S$).

Therefore, $L^\perp \cap S_2$ contains a point $p \notin S$. Using arguments of the second step, we show that the intersection of S with the plane $\langle L, p \rangle$ is a line. This line intersects L which contradicts the inclusion $L \subset S_1 \setminus S$. Hence this case is not realized. □

4.8 Apartments

4.8.1 *Apartments in polar Grassmannians*

Let $\Pi = (P, \mathcal{L})$ be a polar space of rank n and $B = \{p_1, \ldots, p_{2n}\}$ be a frame of Π. Denote by $\mathcal{A}_k$ the associated apartment of $\mathcal{G}_k(\Pi)$, $k \in \{0, 1, \ldots, n-1\}$. It consists of all k-dimensional singular subspaces spanned by subsets of the frame B; in particular, $\mathcal{A}_0$ coincides with B.

Proposition 4.26. $|\mathcal{A}_k| = 2^{k+1} \binom{n}{k+1}$.

Proof. The apartment $\mathcal{A}_k$ consists of all k-dimensional singular subspaces $\langle p_{i_1}, \ldots, p_{i_{k+1}} \rangle$ such that

$$\{i_1, \ldots, i_{k+1}\} \cap \{\sigma(i_1), \ldots, \sigma(i_{k+1})\} = \emptyset.$$

There are $2n$ possibilities to choose p_{i_1}, then p_{i_2} can be chosen in $2n - 2$ ways and so on. Since the order of the points is not taken into account, $\mathcal{A}_k$ consists of

$$\frac{2n \cdot (2n - 2) \cdots (2n - 2k)}{(k + 1)!} = 2^{k+1} \binom{n}{k + 1}$$

elements. □

Exercise 4.13. Show that for any $S, U \in \mathcal{G}_k(\Pi)$ the intersection of all apartments of $\mathcal{G}_k(\Pi)$ containing S and U coincides with $\{S, U\}$.

Consider the set

$$J = \{1, \ldots, n, -1, \ldots, -n\}.$$

Recall that a subset $X \subset J$ is called *singular* if

$$j \in X \implies -j \notin X.$$

We define the *polar Johnson graph* $PJ(n, k)$, $k \in \{0, 1, \ldots, n-1\}$, as the graph whose vertex set is formed by all singular subsets consisting of $k+1$ elements. Two such subsets are connected by an edge if their intersection consists of k elements, and in the case when $k < n-1$, we also require that their union is singular. Then $PJ(n, n-1)$ coincides with the n-dimensional hypercube graph H_n. By the definition, a subset of $P = \mathcal{G}_0(\Pi)$ is a frame of Π if the restriction of $\Gamma_0(\Pi)$ to this subset is isomorphic to $PJ(n, 0)$.

For every subset $\mathcal{X} \subset \mathcal{G}_k(\Pi)$ we denote by $\Gamma(\mathcal{X})$ the restriction of the Grassmann graph $\Gamma_k(\Pi)$ to $\mathcal{X}$. Then $\Gamma(\mathcal{A}_k)$ is isomorphic to the polar Johnson graph $PJ(n, k)$. In the case when $0 < k < n-1$, the maximal cliques of $\Gamma(\mathcal{A}_k)$ are the intersections of $\mathcal{A}_k$ with the stars

$$[M, N]_k, \quad M \in \mathcal{A}_{k-1}, N \in \mathcal{A}_{n-1},$$

and the tops

$$\langle T]_k, \quad T \in \mathcal{A}_{k+1}$$

(if $k = n-2$ then the first possibility is not realized); every maximal clique of $\Gamma(\mathcal{A}_k)$ is an independent subset spanning a maximal singular space of $\mathfrak{G}_k(\Pi)$. The maximal cliques of $\Gamma(\mathcal{A}_{n-1})$ are pairs of adjacent elements; for every $S \in \mathcal{A}_{n-1}$ there are precisely n elements $S_1, \ldots, S_n \in \mathcal{A}_{n-1}$ adjacent with S and

$$S \cap S_1, \ldots, S \cap S_n$$

form a base of the top $\langle S]_{n-2}$.

The following example shows that the Grassmann space $\mathfrak{G}_k(\Pi)$ does not need to be spanned by apartments of $\mathcal{G}_k(\Pi)$.

Example 4.12. Suppose that $\Pi = \Pi_\Omega$, where Ω is a non-degenerate trace-valued reflexive form defined on an n-dimensional vector space V. The Witt index of Ω is assumed to be less than $\frac{n}{2}$. Every frame of Π spans a proper subspace U in Π_V. The associated apartment of $\mathcal{G}_k(\Pi)$ is contained in the subspace

$$\{ S \in \mathcal{G}_k(\Pi) \ : \ S \subset U \}$$

By Lemma 4.7, this is a proper subspace of $\mathfrak{G}_k(\Pi)$.

Let M be an m-dimensional singular subspace of Π. Consider the parabolic subspace $[M\rangle_k$, $m < k$. We take any frame of Π which contains a subset spanning M; the intersection of $[M\rangle_k$ with the associated apartment of $\mathcal{G}_k(\Pi)$ is called an *apartment* of our parabolic subspace. By Lemma 4.4, the natural collineation of $[M\rangle_k$ to the Grassmann space of index $k - m - 1$ associated with the polar space $[M\rangle_{m+1}$ (Example 4.6) establishes a one-to-one correspondence between apartments. The restrictions of the Grassmann graph $\Gamma_k(\Pi)$ to apartments of $[M\rangle_k$ are isomorphic to the polar Johnson graph $PJ(n - m - 1, k - m - 1)$; in particular, the restriction of $\Gamma_{n-1}(\Pi)$ to an apartment of $[M\rangle_{n-1}$ is isomorphic to the $(n - m - 1)$-dimensional hypercube graph.

Consider a subset $\mathcal{X} \subset \mathcal{G}_{n-1}(\Pi)$ and for every $S \in \mathcal{X}$ denote by $\mathcal{X}_S$ the set of all elements of $\mathcal{X}$ adjacent with S. We say that $\mathcal{X}$ is *locally independent* if for every $S \in \mathcal{X}$ the set of all $S \cap U$, $U \in \mathcal{X}_S$, is an independent subset of $\langle S]_{n-2}$. Apartments of $\mathcal{G}_{n-1}(\Pi)$ and apartments in parabolic subspaces of $\mathfrak{G}_{n-1}(\Pi)$ are locally independent.

Theorem 4.14 ([Cooperstein, Kasikova and Shult (2005)]). *Let $\mathcal{X}$ be a locally independent subset of $\mathcal{G}_{n-1}(\Pi)$ such that $\Gamma(\mathcal{X})$ is isomorphic to the m-dimensional hypercube graph, $2 \le m \le n$. If $m = n$ then $\mathcal{X}$ is an apartment of $\mathcal{G}_{n-1}(\Pi)$; in the case when $m < n$, this is an apartment in a parabolic subspace of $\mathfrak{G}_{n-1}(\Pi)$.*

The proof is similar to the proof of Theorem 3.8.

Proof. Let $S \in \mathcal{X}$. There are precisely m elements of $\mathcal{X}$ adjacent with S; we denote them by $S_1, \ldots, S_m$. Then

$$M(S) := S \cap S_1 \cap \cdots \cap S_m$$

is an $(n-m-1)$-dimensional singular subspace and for every $i \in \{1, \ldots, m\}$ the singular subspace

$$X_i := \bigcap_{j \ne i}(S \cap S_j)$$

is $(n - m)$-dimensional. The set formed by all X_i will be denoted by $\mathcal{B}_0(S)$; this is a base of $[M(S), S]_{n-m}$ if $m < n$ and a base of the singular subspace S if $m = n$.

Let U be an element of $\mathcal{X}$ adjacent with S. For every $S' \in \mathcal{X}$ adjacent with S and distinct from U there exists unique $U' \in \mathcal{X}$ adjacent with both S', U and non-adjacent with S (this follows directly from the fact that $\Gamma(\mathcal{X})$ is isomorphic to H_m). It is not difficult to see that

$$S' \cap S \cap U = U' \cap S \cap U.$$

This implies the following:

(1) $M(S) = M(U)$,
(2) every $X \in \mathcal{B}_0(S)$ satisfying $X \subset S \cap U$ belongs to $\mathcal{B}_0(U)$; therefore, $\mathcal{B}_0(S) \cap \mathcal{B}_0(U)$ consists of $m - 1$ elements and there is a unique element of $\mathcal{B}_0(S)$ which does not belong to $\mathcal{B}_0(U)$.

By connectedness of polar Grassmann spaces, we have
$$M(S) = M(U) \quad \forall\, S, U \in \mathcal{X};$$
in what follows this $(n - m - 1)$-dimensional singular subspace will be denoted by M.

For every $S \in \mathcal{X}$ we denote by $\mathcal{B}(S)$ the union of all $\mathcal{B}_0(U)$ such that $U \in \mathcal{X}$ is adjacent with S. Then $\mathcal{B}(S)$ consists of $2m$ elements; moreover, this is a frame of the polar space $[M\rangle_{n-m}$ if $m < n$ and a frame of Π if $m = n$. Now, we show that $\mathcal{B}(S)$ coincides with $\mathcal{B}(U)$ if $S, U \in \mathcal{X}$ are adjacent.

If $X \in \mathcal{B}_0(S)$ then $X \in \mathcal{B}(U)$ (since $\mathcal{B}_0(S)$ is contained in $\mathcal{B}(U)$ by the definition). Consider the case when $X \in \mathcal{B}_0(S') \setminus \mathcal{B}_0(S)$ for a certain $S' \in \mathcal{X} \setminus \{U\}$ adjacent with S. We take $U' \in \mathcal{X}$ adjacent with both S' and U. Then $S \cap S'$ and $U' \cap S'$ are distinct hyperplanes of S'; since X is not contained in $S \cap S'$ ($X \notin \mathcal{B}_0(S)$), we have $X \subset U' \cap S'$ which means that $X \in \mathcal{B}_0(U') \subset \mathcal{B}(U)$.

By connectedness,
$$\mathcal{B}(S) = \mathcal{B}(U) \quad \forall\, S, U \in \mathcal{X};$$
denote this set by $\mathcal{B}$. In the case when $m = n$, this is a frame of Π and $\mathcal{X}$ is the associated apartment of $\mathcal{G}_{n-1}(\Pi)$. If $m < n$ then $\mathcal{B}$ is a frame of the polar space $[M\rangle_{n-m}$ and $\mathcal{X}$ is the associated apartment of the parabolic subspace $[M\rangle_{n-1}$. $\qquad\square$

Remark 4.15. In the general case, the Grassmann space $\mathfrak{G}_{n-1}(\Pi)$ and its parabolic subspaces are not spanned by apartments. By this reason, we cannot use Theorem 4.14 to prove an analogue of Theorem 3.1. However, in some special cases this is possible, see [Cooperstein, Kasikova and Shult (2005)].

4.8.2 *Apartments in half-spin Grassmannians*

Now suppose that Π is a polar space of type D_n and $n \geq 4$. Let B be a frame of Π and $\mathcal{A}_{n-1}$ be the associated apartment of $\mathcal{G}_{n-1}(\Pi)$. Then
$$\mathcal{A}_\delta := \mathcal{A}_{n-1} \cap \mathcal{G}_\delta(\Pi), \quad \delta \in \{+, -\},$$

is the apartment of the half-spin Grassmannian $\mathcal{G}_\delta(\Pi)$ associated with the frame B.

Proposition 4.27. $|\mathcal{A}_\delta| = 2^{n-1}$.

Proof.   This follows from Proposition 4.26, since $|\mathcal{A}_+| = |\mathcal{A}_-|$ and $\mathcal{A}_{n-1}$ is the disjoint union of $\mathcal{A}_+$ and $\mathcal{A}_-$. $\square$

Exercise 4.14. Show that for any $S, U \in \mathcal{G}_\delta(\Pi)$ the intersection of all apartments of $\mathcal{G}_\delta(\Pi)$ containing S and U coincides with $\{S, U\}$.

The vertex set of the hypercube graph H_n (the set of all singular subsets consisting of n elements) can be decomposed in two disjoint subsets $\mathcal{J}_+$ and $\mathcal{J}_-$ (Example 2.2). The distance between two vertices is odd if and only if one of them belongs to $\mathcal{J}_+$ and the other belongs to $\mathcal{J}_-$. The distance between any two vertices from $\mathcal{J}_\delta$, $\delta \in \{+, -\}$, is even. The *half-cube graph* $\frac{1}{2}H_n$ is the graph whose vertex set is $\mathcal{J}_\delta$ and whose edges are pairs of vertices at the distance 2 (this construction does not depend on $\delta \in \{+, -\}$). Note that $\frac{1}{2}H_4$ is isomorphic to $PJ(4, 0)$.

If $\mathcal{X} \subset \mathcal{G}_\delta(\Pi)$ then we write $\Gamma(\mathcal{X})$ for the restriction of the Grassmann graph $\Gamma_\delta(\Pi)$ to the set $\mathcal{X}$. It is clear that $\Gamma(\mathcal{A}_\delta)$ is isomorphic to the half-cube graph $\frac{1}{2}H_n$. The maximal cliques of $\Gamma(\mathcal{A}_\delta)$ are the intersections of $\mathcal{A}_\delta$ with the stars

$$[S\rangle_\delta, \;\; S \in \mathcal{A}_{n-4},$$

and the special subspaces

$$[U]_\delta, \;\; U \in \mathcal{G}_{-\delta}(\Pi);$$

they will be called *stars* and *special subsets* of $\mathcal{A}_\delta$ (respectively). Every maximal clique of $\Gamma(\mathcal{A}_\delta)$ is an independent subset spanning a maximal singular space of $\mathfrak{G}_\delta(\Pi)$.

Let M be an m-dimensional singular subspace of Π and $m < n - 4$. Consider the parabolic subspace $[M\rangle_\delta$ of the half-spin Grassmann space $\mathfrak{G}_\delta(\Pi)$ (Example 4.8). The definition of apartments in $[M\rangle_\delta$ is standard. We identify every $U \in [M\rangle_\delta$ with the subspace $[M, U]_{m+1}$ in the polar space $[M\rangle_{m+1}$. In the present case, $[M\rangle_{m+1}$ is a polar space of type D_{n-m-1} and this correspondence is a collineation of $[M\rangle_\delta$ to one of the half-spin Grassmann spaces of $[M\rangle_{m+1}$. By Lemma 4.4, this collineation establishes a one-to-one correspondence between apartments. The restrictions of the Grassmann graph $\Gamma_\delta(\Pi)$ to apartments of $[M\rangle_\delta$ are isomorphic to the half-cube graph $\frac{1}{2}H_{n-m-1}$.

Theorem 4.15 ([Cooperstein, Kasikova and Shult (2005)]). *Let* Π *be a polar space of type* D_n, $n \geq 4$, *and* $\mathcal{X}$ *be a subset of* $\mathcal{G}_\delta(\Pi)$, $\delta \in \{+, -\}$, *such that* $\Gamma(\mathcal{X})$ *is isomorphic to the half-cube graph* $\frac{1}{2}H_m$, $4 \leq m \leq n$, *and every maximal clique of* $\Gamma(\mathcal{X})$ *is an independent subset of the half-spin Grassmann space* $\mathfrak{G}_\delta(\Pi)$. *If* $m = n$ *then* $\mathcal{X}$ *is an apartment of* $\mathcal{G}_\delta(\Pi)$; *in the case when* $m < n$, *this is an apartment in a parabolic subspace of* $\mathfrak{G}_\delta(\Pi)$.

If Π is of type D_4 then the half-spin Grassmann spaces of Π are polar spaces of type D_4 and every apartment of $\mathcal{G}_\delta(\Pi)$, $\delta \in \{+, -\}$, is a frame of $\mathfrak{G}_\delta(\Pi)$. Since frames are independent subsets of polar spaces, Theorem 4.15 gives the following.

Corollary 4.4. *If* Π *is a polar space of type* D_4 *then the family of all frames of the polar space* $\mathfrak{G}_\delta(\Pi)$, $\delta \in \{+, -\}$, *coincides with the family of all apartments of* $\mathcal{G}_\delta(\Pi)$.

Proposition 4.28. *If* Π *is a polar space of type* D_n, $n \geq 4$, *then half-spin Grassmann space* $\mathfrak{G}_\delta(\Pi)$, $\delta \in \{+, -\}$, *is spanned by every apartment of* $\mathcal{G}_\delta(\Pi)$.

Proof. In the case when $n = 4$, this follows immediately from Proposition 4.9 and Corollary 4.4. Suppose that $n \geq 5$ and prove the statement induction by n. Let $B = \{p_1, \ldots, p_{2n}\}$ be a frame of Π and $\mathcal{A}$ be the associated apartment of $\mathcal{G}_\delta(\Pi)$. Denote by $\mathcal{X}$ the subspace of $\mathfrak{G}_\delta(\Pi)$ spanned by $\mathcal{A}$.

Every $[p_i\rangle_1$ is a polar space of type D_{n-1} and $[p_i\rangle_\delta$ can be identified with one of the half-spin Grassmann spaces of this polar space. Then $\mathcal{A} \cap [p_i\rangle_\delta$ is the apartment of $[p_i\rangle_\delta$ associated with the frame of $[p_i\rangle_1$ formed by all lines $p_i p_j$ with $j \neq i, \sigma(i)$. By the inductive hypothesis, $[p_i\rangle_\delta$ is spanned by $\mathcal{A} \cap [p_i\rangle_\delta$ and we have

$$[p_i\rangle_\delta \subset \mathcal{X}$$

for every i. Now consider $S \in \mathcal{G}_\delta(\Pi)$ such that $p_i \notin S$ for all i. We take any $U \in \mathcal{G}_{-\delta}(\Pi)$ intersecting S in an $(n - 2)$-dimensional subspace and define

$$S_i := \langle p_i, U \cap p_i^\perp \rangle$$

for every i. Then S_i belongs to $[p_i\rangle_\delta$ or coincides with U; the second possibility is realized if $p_i \in U$. We denote by I the set of all i such that $p_i \notin U$; it is clear that $I \neq \emptyset$. The intersection of all $U \cap p_i^\perp$, $i \in I$, is empty (otherwise, there is a point collinear with all points of the frame B which contradicts Corollary 4.1). This means that the projective space $\langle U]_{n-2}$ is spanned by the set of all $U \cap p_i^\perp$, $i \in I$. Then the special subspace $[U]_\delta$ is spanned by the set of all S_i, $i \in I$. Since $S_i \in \mathcal{X}$ for every $i \in I$, we get $S \in [U]_\delta \subset \mathcal{X}$ which completes our proof. $\qquad\square$

Since every parabolic subspace $[M\rangle_\delta$, $M \in \mathcal{G}_m(\Pi)$, can be identified with one of the half-spin Grassmann spaces of the polar space $[M\rangle_{m+1}$ such that apartments correspond to apartments, Proposition 4.28 guarantees that parabolic subspaces are spanned by apartments. As in Section 3.3, we get the following.

Corollary 4.5. *Every subspace of $\mathfrak{G}_\delta(\Pi)$, $\delta \in \{+, -\}$, isomorphic to the half-spin Grassmann space associated with a polar space of type D_m, $m < n$, is parabolic.*

4.8.3 *Proof of Theorem 4.15*

Let Π' be a polar space of type D_m. We take any frame B of Π' and consider the associated apartments

$$\mathcal{A} \subset \mathcal{G}_{m-1}(\Pi'), \quad \mathcal{A}_\delta \subset \mathcal{G}_\delta(\Pi'), \quad \delta \in \{+, -\}.$$

Let $f : \mathcal{A}_+ \to \mathcal{X}$ be an isomorphism of $\Gamma(\mathcal{A}_+)$ to $\Gamma(\mathcal{X})$.

Every maximal clique $\mathcal{Z}$ of $\Gamma(\mathcal{X})$ is contained in precisely one maximal clique of $\Gamma_\delta(\Pi)$ (since $\mathcal{Z}$ is an independent subset of $\mathfrak{G}_\delta(\Pi)$ containing more than 3 elements and the intersection of two distinct maximal cliques of $\Gamma_\delta(\Pi)$ is not greater than a plane); we say that $\mathcal{Z}$ is a *star* of $\mathcal{X}$ or a *special subset* of $\mathcal{X}$ if the maximal clique of $\Gamma_\delta(\Pi)$ containing $\mathcal{Z}$ is a star or a special subspace, respectively. Every special subset $\mathcal{S} \subset \mathcal{A}_+$ consists of m elements. If $f(\mathcal{S})$ is a star then $m = 4$. Therefore, f transfers special subsets of $\mathcal{A}_+$ to special subsets of $\mathcal{X}$ if $m \geq 5$.

Consider the case when $m = 4$. The graphs $\Gamma(\mathcal{A}_+)$ and $\Gamma(\mathcal{X})$ are isomorphic to $PJ(4, 0)$ and every maximal clique of these graphs consists of 4 elements. For any maximal cliques $\mathcal{S}$ and $\mathcal{S}'$ of $\Gamma(\mathcal{A}_+)$ there is a sequence of maximal cliques

$$\mathcal{S} = \mathcal{S}_0, \mathcal{S}_1, \dots, \mathcal{S}_i = \mathcal{S}'$$

such that $|\mathcal{S}_{j-1} \cap \mathcal{S}_j| = 3$ and $\mathcal{S}_{j-1}, \mathcal{S}_j$ are of different types (one of them is a star and the other is a special subset) for every $j \in \{1, \dots, i\}$. If the intersection of two distinct maximal cliques of $\Gamma_\delta(\Pi)$ contains a triangle then these cliques are of different types, and it is not difficult to prove that one of the following possibilities is realized:

(1) special subsets go to special subsets and stars go to stars,

(2) special subsets go to stars and stars go to special subsets.

In the second case, we take any automorphism g of $\Gamma(\mathcal{A}_+)$ satisfying (2) (we leave its construction for the reader); then fg is an isomorphism of $\Gamma(\mathcal{A}_+)$ to $\Gamma(\mathcal{X})$ satisfying (1).

Therefore, we can assume that f transfers special subsets to special subsets in all cases. Then for every $U \in \mathcal{A}_-$ there exists $U' \in \mathcal{G}_{-\delta}(\Pi)$ such that

$$f(\mathcal{A}_+ \cap [U]_+) = \mathcal{X} \cap [U']_\delta.$$

We define $f(U) := U'$ and denote by $\mathcal{Y}$ the subset of $\mathcal{G}_{-\delta}(\Pi)$ consisting of all $f(U)$, $U \in \mathcal{A}_-$. So, f is extended to a bijection of $\mathcal{A}$ to $\mathcal{X} \cup \mathcal{Y}$. This is an isomorphism of $\Gamma(\mathcal{A})$ to $\Gamma(\mathcal{X} \cup \mathcal{Y})$; indeed, $S \in \mathcal{X}$ and $U \in \mathcal{Y}$ are adjacent (as elements of $\mathcal{G}_{n-1}(\Pi)$) if and only if $S \in [U]_\delta$, and the latter is possible only in the case when $f^{-1}(S) \in [f^{-1}(U)]_+$.

So, $\Gamma(\mathcal{X} \cup \mathcal{Y})$ is isomorphic to the m-dimensional hypercube graph and we want to show that $\mathcal{X} \cup \mathcal{Y}$ is a locally independent subset of $\mathcal{G}_{n-1}(\Pi)$.

Let $S \in \mathcal{X} \cup \mathcal{Y}$. As in the proof of Theorem 4.14, we consider

$$S_1, \ldots, S_m \in \mathcal{X} \cup \mathcal{Y}$$

adjacent with S (all S_i belong to $\mathcal{Y}$ if $S \in \mathcal{X}$, and they are elements of $\mathcal{X}$ if $S \in \mathcal{Y}$). The dimension of the subspace

$$M(S) := S \cap S_1 \cap \cdots \cap S_m$$

is not less than $n - m - 1$. In the case when $S \in \mathcal{Y}$,

$$\mathcal{X} \cap [S]_\delta = \{S_1, \ldots, S_m\}$$

is an independent subset of $[S]_\delta$. This implies that all $S \cap S_i$ form an independent subset of $\langle S \rangle_{n-2}$ and the subspace $M(S)$ is $(n-m-1)$-dimensional.

Now, suppose that $S \in \mathcal{X}$. We take any $U \in \mathcal{Y}$ adjacent with S. For every $S' \in \mathcal{Y} \setminus \{U\}$ adjacent with S there exists unique $U' \in \mathcal{X}$ adjacent with both S' and U. As in the proof of Theorem 4.14, we have

$$S' \cap S \cap U = U' \cap S \cap U.$$

This implies that

$$M(S) = M(U).$$

Hence $M(S)$ is $(n - m - 1)$-dimensional which guarantees that all $S \cap S_i$ form an independent subset of $\langle S \rangle_{n-2}$.

Therefore, $\mathcal{X} \cup \mathcal{Y}$ is locally independent. By Theorem 4.14, this is an apartment in $\mathcal{G}_{n-1}(\Pi)$ (if $m = n$) or an apartment in a parabolic subspace of $\mathfrak{G}_{n-1}(\Pi)$ (if $m < n$). This gives the claim.

Remark 4.16. Our proof of Theorem 4.15 is different from the original proof given in [Cooperstein, Kasikova and Shult (2005)].

4.9 Apartments preserving mappings

In this section, we suppose that $\Pi = (P, \mathcal{L})$ and $\Pi' = (P', \mathcal{L}')$ are polar spaces of same type X_n, $\mathsf{X} \in \{\mathsf{C}, \mathsf{D}\}$ and $n \geq 3$; if $\mathsf{X} = \mathsf{D}$ then we require that $n \geq 4$.

4.9.1 *Apartments preserving bijections*

The collineations of $\mathfrak{G}_k(\Pi)$ to $\mathfrak{G}_k(\Pi')$ induced by collineations of Π to Π' are apartments preserving. If our polar spaces are of type D_n then the same holds for the collineations of $\mathfrak{G}_\delta(\Pi)$ to $\mathfrak{G}_\gamma(\Pi')$, $\delta, \gamma \in \{+, -\}$, induced by collineations of Π to Π'.

In the case when Π and Π' are polar spaces of type D_4, we have also the following two types of additional collineations:

- the collineations of $\mathfrak{G}_1(\Pi)$ to $\mathfrak{G}_1(\Pi')$ induced by collineations of Π to the half-spin Grassmann spaces of Π',
- the collineations of $\mathfrak{G}_\delta(\Pi')$ to $\mathfrak{G}_\gamma(\Pi')$, $\delta, \gamma \in \{+, -\}$, induced by collineations of Π to $\mathfrak{G}_{-\gamma}(\Pi')$.

By Corollary 4.4, these collineations are apartments preserving.

Theorem 4.16 (M. Pankov). *Every apartments preserving bijection of* $\mathcal{G}_k(\Pi)$ *to* $\mathcal{G}_k(\Pi')$ *is a collineation of* $\mathfrak{G}_k(\Pi)$ *to* $\mathfrak{G}_k(\Pi')$ *(it is a collinearion of* Π *to* Π' *if* $k = 0$*). If our polar spaces are of type* D_n*,* $n \geq 4$*, then every apartments preserving bijection of* $\mathcal{G}_\delta(\Pi)$ *to* $\mathcal{G}_\gamma(\Pi')$*,* $\delta, \gamma \in \{+, -\}$*, is a collineation of* $\mathfrak{G}_\delta(\Pi')$ *to* $\mathfrak{G}_\gamma(\Pi')$*.*

Remark 4.17. This result was proved in [Pankov 3 (2007)]; for some partial cases it was established in [Pankov 4 (2004); Pankov 2 (2007)].

The proof is based on elementary properties of maximal inexact and complement subsets; the adjacency relation will be characterized in terms of complement subsets. However, in opposite to Grassmannians of finite-dimensional vector spaces, certain polar Grassmanians contain two different types of complement subsets; this makes our proof more complicated.

All apartments preserving mappings will be described in Subsection 4.9.7.

4.9.2   Inexact subsets of polar Grassmannians

Let $B = \{p_1, \ldots, p_{2n}\}$ be a frame of Π and $\mathcal{A}$ be the associated apartment of $\mathcal{G}_k(\Pi)$. Recall that $\mathcal{A}$ consists of all k-dimensional singular subspaces

$$\langle p_{i_1}, \ldots, p_{i_{k+1}} \rangle,$$

where

$$\{i_1, \ldots, i_{k+1}\} \cap \{\sigma(i_1), \ldots, \sigma(i_{k+1})\} = \emptyset.$$

If $k = n - 1$ then every element of $\mathcal{A}$ contains precisely one of the points p_i or $p_{\sigma(i)}$ for each i.

We write $\mathcal{A}(+i)$ and $\mathcal{A}(-i)$ for the sets consisting of all elements of $\mathcal{A}$ which contain p_i and do not contain p_i, respectively. Note that

$$\mathcal{A}(+i) = \mathcal{A}(-\sigma(i))$$

if $k = n - 1$. For any $i_1, \ldots, i_s$ and $j_1, \ldots, j_u$ belonging to $\{1, \ldots, 2n\}$ we define

$$\mathcal{A}(+i_1, \ldots, +i_s, -j_1, \ldots, -j_u)$$

as the intersection

$$\mathcal{A}(+i_1) \cap \cdots \cap \mathcal{A}(+i_s) \cap \mathcal{A}(-j_1) \cap \cdots \cap \mathcal{A}(-j_u).$$

Let $\mathcal{R} \subset \mathcal{A}$. We say that $\mathcal{R}$ is *exact* if there is only one apartment of $\mathcal{G}_k(\Pi)$ containing $\mathcal{R}$; otherwise, $\mathcal{R}$ is said to be *inexact*. If $\mathcal{R} \cap \mathcal{A}(+i)$ is not empty then we define $S_i(\mathcal{R})$ as the intersection of all elements of $\mathcal{R}$ containing p_i; we will write $S_i(\mathcal{R}) = \emptyset$ in the case when the intersection of $\mathcal{R}$ and $\mathcal{A}(+i)$ is empty. If

$$S_i(\mathcal{R}) = p_i$$

for all i then the subset $\mathcal{R}$ is exact. However, in contrast to Grassmannians of finite-dimensional vector space, the converse fails.

Lemma 4.17. *Let $\mathcal{R} \subset \mathcal{A}$. If there exist distinct i, j such that*

$$p_j \in S_i(\mathcal{R}) \quad \text{and} \quad p_{\sigma(i)} \in S_{\sigma(j)}(\mathcal{R}).$$

Then $\mathcal{R}$ is inexact.

Proof.   It is clear that $j \neq \sigma(i)$. On the line $p_i p_j$ we choose a point p_i' distinct from p_i and p_j. The point p_i' is collinear with a unique point of the line $p_{\sigma(i)} p_{\sigma(j)}$; denote this point by $p_{\sigma(j)}'$. Then

$$(B \setminus \{p_i, p_{\sigma(j)}\}) \cup \{p_i', p_{\sigma(j)}'\}$$

is a frame of Π. This frame defines a new apartment of $\mathcal{G}_k(\Pi)$ containing the subset $\mathcal{R}$. $\square$

Remark 4.18. In particular, we established that

$$B \setminus \{p_i, p_j\}, \quad j \neq i, \sigma(i)$$

is an inexact subset of B (frames are apartments in the Grassmannian $\mathcal{G}_0(\Pi) = P$).

Lemma 4.18. *If Π is a polar space of type C_n then for each $i \in \{1, \ldots, 2n\}$ there exists a point $p_i' \neq p_i, p_{\sigma(i)}$ such that*

$$(B \setminus \{p_i\}) \cup \{p_i'\} \quad and \quad (B \setminus \{p_{\sigma(i)}\}) \cup \{p_i'\}$$

are frames of Π.

Proof.   Let S and U be disjoint $(n-2)$-dimensional singular subspaces spanned by subsets of $B \setminus \{p_i, p_{\sigma(i)}\}$. Since Π is a polar space of type C_n, there exists a maximal singular subspace

$$S' \neq \langle S, p_i \rangle, \langle S, p_{\sigma(i)} \rangle$$

containing S. Lemma 4.2 implies the existence of a point $p \in S'$ satisfying $p \perp U$. This point does not belong to S (for every point $q \in S$ we have $q \not\perp U$). Hence p is non-collinear with p_i and $p_{\sigma(i)}$. Since $p \perp (S \cup U)$ and

$$B \setminus \{p_i, p_{\sigma(i)}\} \subset S \cup U,$$

p is collinear with all points of $B \setminus \{p_i, p_{\sigma(i)}\}$. The point $p_i' := p$ is as required. $\square$

Lemma 4.19. *If S and U are singular subspaces spanned by subsets of B then $S^\perp \cap U$ is spanned by a subset of B.*

Proof.   The case when $S \perp U$ is trivial. Suppose that $S \not\perp U$ and denote by X the set of all points from $U \cap B$ which are not contained in $S^\perp$. The singular subspace spanned by $(U \cap B) \setminus X$ is contained in $S^\perp \cap U$. If this subspace does not coincide with $S^\perp \cap U$ then $S^\perp$ has a non-empty intersection with $\langle X \rangle$. Let p be a point belonging to this intersection. There exists $p_i \in X$ such that

$$(X \setminus \{p_i\}) \cup \{p\} \tag{4.6}$$

is a base of $\langle X \rangle$. It is clear that $p_{\sigma(i)}$ belongs to S and it is collinear with all points of the base (4.6). Then $p_{\sigma(i)} \perp \langle X \rangle$ which contradicts $p_i \in \langle X \rangle$. Thus $S^\perp$ does not intersect $\langle X \rangle$ and $S^\perp \cap U$ is spanned by $(U \cap B) \setminus X$. $\square$

Proposition 4.29. *If $k = n - 1$ then $\mathcal{A}(-i)$ is inexact, but this inexact subset is not maximal. For every $k \in \{0, \ldots, n-2\}$ the following assertions are fulfilled:*

(1) $\mathcal{A}(-i)$ *is a maximal inexact subset if Π is of type C_n,*
(2) $\mathcal{A}(-i)$ *is exact if Π is of type D_n.*

Proof. Suppose that $k = n - 1$. In this case, the subset $\mathcal{A}(-i)$ coincides with $\mathcal{A}(+\sigma(i))$. Let us take any U belonging to $\mathcal{A} \setminus \mathcal{A}(-i) = \mathcal{A}(+i)$. Then

$$S_i(\mathcal{A}(-i) \cup \{U\}) = U,$$

$$S_{\sigma(i)}(\mathcal{A}(-i) \cup \{U\}) = S_{\sigma(i)}(\mathcal{A}(-i)) = p_{\sigma(i)}.$$

Since for every $j \neq i, \sigma(i)$ the intersection of all elements of $\mathcal{A}(-i)$ containing p_j is the line $p_j p_{\sigma(i)}$, we get

$$S_j(\mathcal{A}(-i) \cup \{U\}) = p_j, \ \ p_j \in U,$$

$$S_j(\mathcal{A}(-i) \cup \{U\}) = p_j p_{\sigma(i)}, \ \ p_j \notin U.$$

In the second case, $p_{\sigma(j)}$ belongs to U; then

$$p_{\sigma(j)} \in S_i(\mathcal{A}(-i) \cup \{U\}) \ \ \text{and} \ \ p_{\sigma(i)} \in S_j(\mathcal{A}(-i) \cup \{U\}).$$

By Lemma 4.17, the subset

$$\mathcal{A}(-i) \cup \{U\}$$

is inexact. This means that $\mathcal{A}(-i)$ is inexact; moreover, this inexact subset is not maximal.

Consider the case when $k \leq n - 2$. We have

$$S_j(\mathcal{A}(-i)) = p_j \ \ \text{for} \ \ j \neq i \tag{4.7}$$

(it is not difficult to choose two elements of $\mathcal{A}(-i)$ whose intersection is precisely the point p_j). Therefore, if $\mathcal{A}(-i)$ is contained in the apartment of $\mathcal{G}_k(\Pi)$ associated with a frame $B' \neq B$ then

$$B' = (B \setminus \{p_i\}) \cup \{p\}$$

and $p \neq p_i$. The point p is collinear with all points of $B \setminus \{p_i, p_{\sigma(i)}\}$ and non-collinear with $p_{\sigma(i)}$. Also we have $p \not\perp p_i$ (otherwise the point p is contained in every maximal singular subspace spanned by p_i and a subset of $B \setminus \{p_i, p_{\sigma(i)}\}$ which implies $p_i = p$, see Proposition 4.6). If S is an $(n-2)$-dimensional singular subspace spanned by a subset of $B \setminus \{p_i, p_{\sigma(i)}\}$ then the points $p, p_i, p_{\sigma(i)}$ define three distinct maximal singular subspaces containing S which is impossible if Π is of type D_n. Thus $\mathcal{A}(-i)$ is exact if Π is of type D_n.

Suppose that Π is a polar space of type C_n. By Lemma 4.18, there is a point $p_i' \neq p_i, p_{\sigma(i)}$ such that

$$(B \setminus \{p_i\}) \cup \{p_i'\}$$

is a frame of Π. The associated apartment of $\mathcal{G}_k(\Pi)$ contains $\mathcal{A}(-i)$. The subset $\mathcal{A}(-i)$ is inexact.

Let U be an element of $\mathcal{A} \setminus \mathcal{A}(-i) = \mathcal{A}(+i)$. Then U is spanned by p_i and some points $p_{i_1}, \ldots, p_{i_k}$. By Lemma 4.19, p_i is the unique point of U collinear with all points $p_{\sigma(i_1)}, \ldots, p_{\sigma(i_k)}$ and (4.7) guarantees that the subset

$$\mathcal{A}(-i) \cup \{U\}$$

is exact. Since this is true for every $U \in \mathcal{A} \setminus \mathcal{A}(-i)$, the inexact subset $\mathcal{A}(-i)$ is maximal. $\qquad \square$

Consider the subset

$$\mathcal{R}_{ij} := \mathcal{A}(+i, +j) \cup \mathcal{A}(+\sigma(i), +\sigma(j)) \cup \mathcal{A}(-i, -\sigma(j)), \quad j \neq i, \sigma(i).$$

It is clear that

$$\mathcal{R}_{ij} = \mathcal{A}(+i, +j) \cup \mathcal{A}(-i) \quad \text{if} \quad k = n - 1$$

and

$$\mathcal{R}_{ij} = B \setminus \{p_i, p_{\sigma(j)}\} \quad \text{if} \quad k = 0.$$

Now we prove the following equality

$$\mathcal{A} \setminus \mathcal{R}_{ij} = \mathcal{A}(+i, -j) \cup \mathcal{A}(+\sigma(j), -\sigma(i)). \tag{4.8}$$

Proof. Let U be an element of $\mathcal{A} \setminus \mathcal{R}_{ij}$. Then U does not belong to $\mathcal{A}(-i, -\sigma(j))$ and we have

$$U \in \mathcal{A}(+i) \quad \text{or} \quad U \in \mathcal{A}(+\sigma(j)).$$

In the first case,

$$U \notin \mathcal{A}(+i, +j) \implies U \in \mathcal{A}(+i, -j).$$

In the second case,

$$U \notin \mathcal{A}(+\sigma(i), +\sigma(j)) \implies U \in \mathcal{A}(+\sigma(j), -\sigma(i)).$$

Therefore,

$$\mathcal{A} \setminus \mathcal{R}_{ij} \subset \mathcal{A}(+i, -j) \cup \mathcal{A}(+\sigma(j), -\sigma(i)).$$

Since $\mathcal{A}(+i, -j)$ and $\mathcal{A}(+\sigma(j), -\sigma(i))$ do not intersect $\mathcal{R}_{ij}$, the inverse inclusion holds. $\qquad\square$

Proposition 4.30. *The subset $\mathcal{R}_{ij}$ is inexact; moreover, it is a maximal inexact subset, except the case when $k = 0$ and Π is of type C_n.*

Proof. Let $k = 0$. Then $\mathcal{R}_{ij} = B \setminus \{p_i, p_{\sigma(j)}\}$ is an inexact subset (Remark 4.18). If Π is of type C_n then $\mathcal{A}(-i) = B \setminus \{p_i\}$ is an inexact subset containing $\mathcal{R}_{ij}$ (Proposition 4.29) and the inexact subset $\mathcal{R}_{ij}$ is not maximal. In the case when Π is of type D_n, the subsets

$$\mathcal{A}(-i) = B \setminus \{p_i\} \text{ and } \mathcal{A}(-\sigma(j)) = B \setminus \{p_{\sigma(j)}\}$$

are exact (Proposition 4.29) which means that $\mathcal{R}_{ij}$ is a maximal inexact subset.

Suppose that $k \geq 1$. Since

$$S_i(\mathcal{R}_{ij}) = p_i p_j \text{ and } S_{\sigma(j)}(\mathcal{R}_{ij}) = p_{\sigma(j)} p_{\sigma(i)},$$

Lemma 4.17 implies that the subset $\mathcal{R}_{ij}$ is inexact.

For every $l \neq i, \sigma(j)$ we can choose two elements of $\mathcal{R}_{ij}$ whose intersection is p_l (we leave the details for the reader). This means that

$$S_l(\mathcal{R}_{ij}) = p_l \text{ if } l \neq i, \sigma(j).$$

By (4.8), for every U belonging to $\mathcal{A} \setminus \mathcal{R}_{ij}$ one of the following possibilities is realized:

- $U \in \mathcal{A}(+i, -j)$ intersects $S_i(\mathcal{R}_{ij}) = p_i p_j$ precisely in the point p_i; since $p_{\sigma(j)}$ is the unique point on the line $S_{\sigma(j)}(\mathcal{R}_{ij}) = p_{\sigma(j)} p_{\sigma(i)}$ collinear with p_i, the set $\mathcal{R}_{ij} \cup \{U\}$ is exact.
- $U \in \mathcal{A}(+\sigma(j), -\sigma(i))$ intersects $S_{\sigma(j)}(\mathcal{R}_{ij}) = p_{\sigma(j)} p_{\sigma(i)}$ precisely in the point $p_{\sigma(j)}$; since p_i is the unique point on the line $S_i(\mathcal{R}_{ij}) = p_i p_j$ collinear with $p_{\sigma(j)}$, the set $\mathcal{R}_{ij} \cup \{U\}$ is exact.

Therefore, $\mathcal{R}_{ij} \cup \{U\}$ is exact for each $U \in \mathcal{A} \setminus \mathcal{R}_{ij}$ and the inexact subset $\mathcal{R}_{ij}$ is maximal. $\qquad\square$

If the set $\mathcal{A}(-i)$ is inexact (see Proposition 4.29) then it will be called an *inexact subset of first type*. Similarly, if $\mathcal{R}_{ij}$ is inexact (see Proposition 4.30) then it is said to be an *inexact subset of second type*.

Proposition 4.31. *Every maximal inexact subset is of first type or of second type. In particular, the following assertions are fulfilled:*

(1) *if $k = n - 1$ then each maximal inexact subset is of second type,*
(2) *if $k \in \{1, \ldots, n - 2\}$ and Π is of type C_n then each maximal inexact subset is of first type or of second type,*
(3) *if $k \in \{0, \ldots, n - 2\}$ and Π is of type D_n then each maximal inexact subset is of second type,*
(4) *if $k = 0$ and Π is of type C_n then each maximal inexact subset is of first type.*

Proof.   Let $k = 0$. The statement is trivial if Π is of type C_n.

Suppose that Π is of type D_n. We need to show that $B \setminus \{p_i, p_{\sigma(i)}\}$ is exact. Assume that there exist points $p'_i, p'_{\sigma(i)}$ such that

$$(B \setminus \{p_i, p_{\sigma(i)}\}) \cup \{p'_i, p'_{\sigma(i)}\}$$

is a frame. For every $(n - 2)$-dimensional singular subspace S spanned by a subset of $B \setminus \{p_i, p_{\sigma(i)}\}$ the pairs of subspaces

$$\langle S, p_i \rangle, \langle S, p_{\sigma(i)} \rangle \text{ and } \langle S, p'_i \rangle, \langle S, p'_{\sigma(i)} \rangle$$

are coincident (since Π is of type D_n). This means that $\{p_i, p_{\sigma(i)}\}$ coincides with $\{p'_i, p'_{\sigma(i)}\}$ (we leave the details for the reader) and $B \setminus \{p_i, p_{\sigma(i)}\}$ is exact.

Let $k \geq 1$ and $\mathcal{R}$ be a maximal inexact subset of $\mathcal{A}$. First, we consider the case when all $S_i(\mathcal{R})$ are non-empty. Denote by I the set of all numbers i such that the dimension of $S_i(\mathcal{R})$ is greater than zero. The subset $\mathcal{R}$ is inexact and I is non-empty. Suppose that for certain $l \in I$ the subspace $S_l(\mathcal{R})$ is spanned by $p_l, p_{j_1}, \ldots, p_{j_u}$ and the subspaces

$$M_1 := S_{\sigma(j_1)}(\mathcal{R}), \ldots, M_u := S_{\sigma(j_u)}(\mathcal{R})$$

do not contain the point $p_{\sigma(l)}$. Then p_l is collinear with all points of the subspaces $M_1, \ldots, M_u$. Since

$$p_{j_1} \not\perp M_1, \ldots, p_{j_u} \not\perp M_u,$$

p_l is the unique point of $S_l(\mathcal{R})$ collinear with all points of $M_1, \ldots, M_u$ (Lemma 4.19). If the same holds for every $l \in I$ then Lemma 4.19 implies that $\mathcal{R}$ is exact. Therefore, there exist $i \in I$ and $j \neq i, \sigma(i)$ such that

$$p_j \in S_i(\mathcal{R}) \text{ and } p_{\sigma(i)} \in S_{\sigma(j)}(\mathcal{R}).$$

Then $\mathcal{R}$ is contained $\mathcal{R}_{ij}$. Indeed, for every $U \in \mathcal{R}$ one of the following possibilities is realized:

- $p_i \in U$ then $U \in \mathcal{A}(+i, +j) \subset \mathcal{R}_{ij}$,
- $p_{\sigma(j)} \in U$ then $U \in \mathcal{A}(+\sigma(i), +\sigma(j)) \subset \mathcal{R}_{ij}$,
- $p_i, p_{\sigma(j)} \notin U$ then $U \in \mathcal{A}(-i, -\sigma(j)) \subset \mathcal{R}_{ij}$.

We have $\mathcal{R} = \mathcal{R}_{ij}$, since the inexact subset $\mathcal{R}$ is maximal.

Now suppose that $S_i(\mathcal{R}) = \emptyset$ for certain i. Then $\mathcal{R}$ is contained in $\mathcal{A}(-i)$ and we get a maximal inexact subset of first type if Π is of type C_n. If the polar space is of type D_n then $\mathcal{A}(-i)$ is exact and one of the following possibilities is realized:

(1) $S_j(\mathcal{R}) = \emptyset$ for certain $j \neq i, \sigma(i)$. Then $\mathcal{R} \subset \mathcal{A}(-j, -i)$ and $\mathcal{R}$ is a proper subset of $\mathcal{R}_{j\,\sigma(i)}$.
(2) All $S_l(\mathcal{R})$, $l \neq i, \sigma(i)$, are non-empty and there exists $j \neq i, \sigma(i)$ such that the subspace $S_j(\mathcal{R})$ contains $p_{\sigma(i)}$. Since every $U \in \mathcal{R} \setminus \mathcal{A}(+j)$ is contained in $\mathcal{A}(-j, -i)$ and every $U \in \mathcal{R} \cap \mathcal{A}(+j)$ belongs to $\mathcal{A}(+j, +\sigma(i))$, we obtain the inclusion
$$\mathcal{R} \subset \mathcal{A}(+j, +\sigma(i)) \cup \mathcal{A}(-j, -i).$$
As in the previous case, $\mathcal{R}$ is a proper subset of $\mathcal{R}_{j\,\sigma(i)}$.
(3) Each $S_j(\mathcal{R})$, $j \neq i, \sigma(i)$ is non-empty and does not contain $p_{\sigma(i)}$.

In the cases (1) and (2), $\mathcal{R}$ is a proper subset of a maximal inexact subset which contradicts our assumption.

Consider the case (3). Since the subset $\mathcal{R}$ is inexact, there exists a frame $B' \neq B$ such that $\mathcal{R}$ is contained in the associated apartment of $\mathcal{G}_k(\Pi)$. We denote by I the set of all numbers $j \neq i, \sigma(i)$ such that the dimension of $S_j(\mathcal{R})$ is greater than zero. If I is empty then
$$B' = (B \setminus \{p_i, p_{\sigma(i)}\}) \cup \{p'_i, p'_{\sigma(i)}\}. \tag{4.9}$$
As in the case when $k = 0$, we get $\{p_i, p_{\sigma(i)}\} = \{p'_i, p'_{\sigma(i)}\}$ which implies that $B = B'$. Therefore, $I \neq \emptyset$.

Let $l \in I$. If $S_l(\mathcal{R})$ is spanned by $p_l, p_{j_1}, \ldots, p_{j_u}$ (by our assumption, these points are distinct from p_i and $p_{\sigma(i)}$) and the subspaces
$$M_1 := S_{\sigma(j_1)}(\mathcal{R}), \ldots, M_u := S_{\sigma(j_u)}(\mathcal{R})$$
do not contain $p_{\sigma(l)}$ then p_l is the unique point of $S_l(\mathcal{R})$ collinear with all points of $M_1, \ldots, M_u$. If the same holds for all elements of I then, by Lemma 4.19, we get (4.9) again. As above, we establish that $\mathcal{R}$ is contained in a certain complement subset $\mathcal{R}_{jm}$, $j, m \neq i, \sigma(i)$. The inclusion
$$\mathcal{R} \subset \mathcal{A}(-i) \cap \mathcal{R}_{jm}$$
means that $\mathcal{R}$ is a proper subset of $\mathcal{R}_{jm}$ which is impossible. $\qquad\square$

4.9.3 *Complement subsets of polar Grassmannians*

Let $\mathcal{A}$ be as in the previous subsection. We say that $\mathcal{R} \subset \mathcal{A}$ is a *complement subset* if $\mathcal{A} \setminus \mathcal{R}$ is a maximal inexact subset. A complement subset is said to be of *first type* or of *second type* if the corresponding maximal inexact subset is of first type or of second type, respectively. If $\mathcal{A}(-i)$ is a maximal inexact subset (this fails in some cases) then the associated complement subset is $\mathcal{A}(+i)$. By (4.8), the maximal inexact subset $\mathcal{R}_{ij}$ ($k > 0$ or Π is of type D_n) gives the complement subset

$$\mathcal{C}_{ij} := \mathcal{A}(+i, -j) \cup \mathcal{A}(+\sigma(j), -\sigma(i))$$

which coincides with $\mathcal{A}(+i, +\sigma(j))$ if $k = n - 1$. Note that

$$\mathcal{C}_{ij} = \mathcal{C}_{\sigma(j)\sigma(i)}.$$

The follows assertions follow immediately from Proposition 4.31:

(1) if $k = n - 1$ then each complement subset is of second type;
(2) if $k \in \{1, \ldots, n - 2\}$ and Π is of type C_n then each complement subset is of first type or of second type;
(3) if $k \in \{0, \ldots, n - 2\}$ and Π is of type D_n then each complement subset is of second type; in the case when $k = 0$, this is a pair of collinear points;
(4) if $k = 0$ and Π is of type C_n then each complement subset is of first type (a single point).

Now we characterize the adjacency and weak adjacency relations of $\mathcal{G}_k(\Pi)$, $k \geq 1$, in terms of complement subsets of second type. In the case when $k = n - 1$ or the polar space is of type D_n, each complement subset is of such type. If S, U are distinct elements of $\mathcal{A}$ then we write $c(S, U)$ for the number of compliment subsets of second type in $\mathcal{A}$ containing both S and U. Let us define

$$\mathrm{m}_c := \max\{ \, c(S, U) \, : \, S, U \in \mathcal{A}, \, S \neq U \, \}.$$

Lemma 4.20. *Let $k \geq 1$ and S, U be distinct elements of $\mathcal{A}$. The following assertions are fulfilled:*

(1) *S, U are adjacent if and only if $c(S, U) = \mathrm{m}_c$,*
(2) *if $k < n - 1$ and $n \neq 4$ or $k \neq 1$ then S, U are weakly adjacent if and only if $c(S, U) = \mathrm{m}_c - 1$.*

Proof.   Let $S, U \in \mathcal{A}$ and m be the dimension of $S \cap U$. If $k = n - 1$ then the complement subset $\mathcal{A}(+i, +j)$ contains both S, U if and only if the line $p_i p_j$ is contained $S \cap U$; thus

$$c(S, U) = \binom{m+1}{2}$$

and we get the claim.

Suppose that $1 \le k \le n - 2$. If the complement subset $\mathcal{C}_{ij} = \mathcal{C}_{\sigma(j)\sigma(i)}$ contains S and U then one of the following possibilities is realized:

(A) the subsets $\mathcal{A}(+i, -j)$ and $\mathcal{A}(+\sigma(j), -\sigma(i))$ both contain S and U;

(B) only one of the subsets $\mathcal{A}(+i, -j)$ or $\mathcal{A}(+\sigma(j), -\sigma(i))$ contains both S, U;

(C) each of the subsets $\mathcal{A}(+i, -j)$ and $\mathcal{A}(+\sigma(j), -\sigma(i))$ contains precisely one of the subspaces.

The case (A). We have

$$S, U \in \mathcal{A}(+i, -j) \cap \mathcal{A}(+\sigma(j), -\sigma(i))$$

if and only if

$$p_i, p_{\sigma(j)} \in S \cap U.$$

Thus there are precisely

$$\binom{m+1}{2}$$

distinct $\mathcal{C}_{ij} = \mathcal{C}_{\sigma(j)\sigma(i)}$ satisfying (A).

The case (B). If S and U belong to $\mathcal{A}(+i, -j)$ then $\mathcal{A}(+\sigma(j), -\sigma(i))$ does not contain the subset $\{S, U\}$. As above, $p_i \in S \cap U$ and there are $m + 1$ possibilities for i. Also we have

$$p_j \notin S \cup U \quad \text{and} \quad p_{\sigma(j)} \notin S \cap U.$$

Hence there are

$$2n - 2(k - m) - 2(m + 1) = 2(n - k - 1)$$

possibilities for j. This means that there exist precisely

$$2(m + 1)(n - k - 1)$$

distinct $\mathcal{C}_{ij} = \mathcal{C}_{\sigma(j)\sigma(i)}$ satisfying (B).

The case (C). Suppose that

$$S \in \mathcal{A}(+i, -j) \setminus \mathcal{A}(+\sigma(j), -\sigma(i)) \quad \text{and} \quad U \in \mathcal{A}(+\sigma(j), -\sigma(i)) \setminus \mathcal{A}(+i, -j).$$

Then $p_{\sigma(j)} \in U$. Since U does not belong to $\mathcal{A}(+i, -j)$ and $p_j \notin U$, we have $p_i \notin U$. Similarly, $p_i \in S$ implies that $p_{\sigma(j)} \notin S$. So,

$$p_i \in S \setminus U \quad \text{and} \quad p_{\sigma(j)} \in U \setminus S.$$

Since $p_{\sigma(i)} \notin U$ and $p_j \notin S$,

$$p_i \perp U \quad \text{and} \quad p_{\sigma(j)} \perp S.$$

The set $B \setminus U$ contains precisely $2(n - k - 1)$ points collinear with all points of U and $B \cap (S \setminus U)$ contains not greater than $n - k - 1$ such points (if $p_l \in B \setminus U$ is collinear with all points of U then the same holds for $p_{\sigma(l)} \in B \setminus U$, but $S \setminus U$ does not contain at least one of these points). Similarly, $U \setminus S$ contains not greater than $n - k - 1$ points of B collinear with all points of S. Therefore, there exist at most

$$l(m) := \min\{(k - m)^2, (n - k - 1)^2\}$$

distinct $\mathcal{C}_{ij} = \mathcal{C}_{\sigma(j)\sigma(i)}$ satisfying (C). In the case when $m = k - 1$ (S and U are adjacent or weakly adjacent), we have

$$B \cap (S \setminus U) = \{p_u\} \quad \text{and} \quad B \cap (U \setminus S) = \{p_v\}.$$

If S and U are adjacent then $v \neq \sigma(u)$ and $\mathcal{C}_{u\sigma(v)} = \mathcal{C}_{v\sigma(u)}$ is the unique compliment subset satisfying (C). If S and U are weakly adjacent then there are no compliment subsets of such kind.

So, we established that

$$c(S, U) \leq \binom{m + 1}{2} + 2(m + 1)(n - k - 1) + l(m);$$

moreover,

$$c(S, U) = \binom{k}{2} + 2k(n - k - 1) + 1$$

if S, U are adjacent and

$$c(S, U) = \binom{k}{2} + 2k(n - k - 1)$$

if they are weakly adjacent. A direct verification shows that

$$\binom{k}{2} + 2k(n - k - 1) \geq \binom{m + 1}{2} + 2(m + 1)(n - k - 1) + l(m)$$

for $m \leq k - 2$ and we get the equality only in the case when $n = 4$ and $k = 1$. $\qquad\square$

Remark 4.19. Suppose that Π is of type C_n and $1 \le k \le n - 2$. In this case, the adjacency and weak adjacency relations can be characterized in terms of maximal inexact subsets of first type. For the intersection of $2n - k - 2$ distinct maximal inexact subsets of first type one of the following possibilities is realized:

- the intersection is empty,
- it consists of $k + 2$ mutually adjacent elements of $\mathcal{A}$,
- it consists of 2 weakly adjacent elements of $\mathcal{A}$.

Lemma 4.21. *Suppose that $1 \le k \le n - 2$ and Π is a polar space of type C_n. Let $\mathcal{R}$ be a complement subset of $\mathcal{A}$. If $\mathcal{R}$ is of first type then there are precisely $2n - 1$ distinct complement subsets of $\mathcal{A}$ disjoint from $\mathcal{R}$. If $\mathcal{R}$ is of second type then there are precisely 3 distinct complement subsets of $\mathcal{A}$ disjoint from $\mathcal{R}$.*

Proof. We take any $l \in \{1, \ldots, 2n\}$ and consider the complement subset $\mathcal{A}(+l)$. There is only one complement subset of first type which does not intersect $\mathcal{A}(+l)$; this is $\mathcal{A}(+\sigma(l))$. If a complement subset

$$\mathcal{C}_{ij} = \mathcal{C}_{\sigma(j)\sigma(i)} = \mathcal{A}(+i, -j) \cup \mathcal{A}(+\sigma(j), -\sigma(i))$$

is disjoint from $\mathcal{A}(+l)$ then

$$i = \sigma(l) \quad \text{or} \quad j = l;$$

in other words, this is $\mathcal{C}_{\sigma(l)m} = \mathcal{C}_{\sigma(m)l}$ and we have $2n - 2$ possibilities for m (since $m \ne l, \sigma(l)$). Therefore, there are precisely $2n - 2 + 1$ distinct complement subsets of $\mathcal{A}$ disjoint from $\mathcal{A}(+l)$.

Now let us fix $i, j \in \{1, \ldots, 2n\}$ such that $j \ne i, \sigma(i)$ and consider the associated complement subset $\mathcal{C}_{ij} = \mathcal{C}_{\sigma(j)\sigma(i)}$. There are only two complement subsets of first type disjoint from it:

$$\mathcal{A}(+\sigma(i)) \quad \text{and} \quad \mathcal{A}(+j).$$

Suppose that $\mathcal{C}_{i'j'}$ does not intersect $\mathcal{C}_{ij}$. Then

$$\mathcal{A}(+i, -j) \cap \mathcal{A}(+i', -j') = \emptyset \implies i' = \sigma(i) \quad \text{or} \quad i' = j \quad \text{or} \quad j' = i.$$

Using the equalities

$$\mathcal{A}(+\sigma(j), -\sigma(i)) \cap \mathcal{A}(+i', -j') = \emptyset,$$
$$\mathcal{A}(+i, -j) \cap \mathcal{A}(+\sigma(j'), -\sigma(i')) = \emptyset,$$
$$\mathcal{A}(+\sigma(j), -\sigma(i)) \cap \mathcal{A}(+\sigma(j'), -\sigma(i')) = \emptyset,$$

we establish that

$$i' = j, \ j' = i \quad \text{or} \quad i' = \sigma(i), j' = \sigma(j).$$

Thus $\mathcal{C}_{ji} = \mathcal{C}_{\sigma(i)\sigma(j)}$ is the unique complement subset of second type disjoint from $\mathcal{C}_{ij} = \mathcal{C}_{\sigma(j)\sigma(i)}$. $\qquad\square$

Proposition 4.32. *Let* $f : \mathcal{G}_k(\Pi) \to \mathcal{G}_k(\Pi')$ *be an apartments preserving mapping and* $k \geq 1$. *Then* f *is adjacency preserving* (*two elements of* $\mathcal{G}_k(\Pi)$ *are adjacent if and only if their images are adjacent*). *In the case when* $k \leq n - 2$, *if* $k \neq 1$ *or our polar spaces are not of type* D_4 *then* f *is weak adjacency preserving* (*two elements of* $\mathcal{G}_k(\Pi)$ *are weakly adjacent if and only if their images are weakly adjacent*).

Proof. First of all, recall that the mapping f is injective. Let S and U be distinct elements of $\mathcal{G}_k(\Pi)$ and $\mathcal{A}$ be an apartment of $\mathcal{G}_k(\Pi)$ containing them. Then $f(S)$ and $f(U)$ belong to the apartment $f(\mathcal{A})$. As in the proof of Proposition 3.6, we show that f transfers inexact subsets of $\mathcal{A}$ to inexact subsets of $f(\mathcal{A})$. Since $\mathcal{A}$ and $f(\mathcal{A})$ have the same number of inexact subsets (Π and Π' are polar spaces of the same type), each inexact subset of $f(\mathcal{A})$ is the image of a certain inexact subset of $\mathcal{A}$. Then an inexact subset of $\mathcal{A}$ is maximal if and only if the same holds for its image. Thus $\mathcal{R}$ is a complement subset of $\mathcal{A}$ if and only if $f(\mathcal{R})$ is a complement subset of $f(\mathcal{A})$.

In the case when $k = n - 1$ or our polar spaces are of type D_n, all complement subsets are of second type; if $1 \leq k \leq n-2$ and the polar spaces are of type C_n then, by Lemma 4.21, the mapping f preserves the types of all maximal inexact and complement subsets. It follows from Lemma 4.20 that S and U are adjacent if and only if $f(S)$ and $f(U)$ are adjacent.

Suppose that $1 \leq k \leq n - 2$. If $n \neq 4$ or $k \neq 1$ then Lemma 4.20 shows that S and U are weakly adjacent if and only if their images are weakly adjacent. In the case when $k = 1$ and the polar spaces are of type C_4, this statement follows immediately from Remark 4.19. $\qquad\square$

Remark 4.20. If the polar spaces are of type D_4 then there exist colineations of $\mathfrak{G}_1(\Pi)$ to $\mathfrak{G}_1(\Pi')$ which do not preserve the weak adjacency relation (Remark 4.13) and the second part of Proposition 4.32 does not hold in this case.

Remark 4.21. If our polar spaces are of type C_n then $\mathcal{G}_k(\Pi)$, $1 \leq k \leq n - 2$, contains complement subsets of both types. Denote by $n_1(n, k)$ and $n_2(n, k)$ the numbers of elements in complement subsets of first and second type, respectively (it is clear that complement subsets of the same type have the same number of elements). Every apartments preserving mapping does not change the types of complement subsets if $n_1(n, k) \neq n_2(n, k)$. However, $n_1(n, k) = n_2(n, k)$ for some pairs n, k and we need Lemma 4.21 to distinguish complement subsets of different types.

4.9.4 *Inexact subsets of half-spin Grassmannians*

Now we suppose that Π is a polar space of type D_n and $n \geq 4$. Let $B = \{p_1, \ldots, p_{2n}\}$ be a frame of Π and $\mathcal{A}$ be the associated apartment of $\mathcal{G}_\delta(\Pi)$, $\delta \in \{+, -\}$. Recall that it consists of all maximal singular subspaces

$$\langle p_{i_1}, \ldots, p_{i_n} \rangle \in \mathcal{G}_\delta(\Pi),$$

where

$$\{i_1, \ldots, i_n\} \cap \{\sigma(i_1), \ldots, \sigma(i_n)\} = \emptyset.$$

Every element of $\mathcal{A}$ contains precisely one of the points p_i or $p_{\sigma(i)}$ for every i. As in Subsection 4.9.2, we define $\mathcal{A}(+i)$, $\mathcal{A}(-i)$,

$$\mathcal{A}(+i_1, \ldots, +i_s, -j_1, \ldots, -j_u)$$

and introduce the concepts of exact, inexact, and complement subsets. For every subset $\mathcal{R} \subset \mathcal{A}$ we denote by $S_i(\mathcal{R})$ the intersection of all elements of $\mathcal{A}(+i) \cap \mathcal{R}$; if this subset is empty then we write $S_i(\mathcal{R}) = \emptyset$.

Proposition 4.33. *If $j \neq i, \sigma(i)$ then*

$$\mathcal{A}(-i) \cup \mathcal{A}(+i, +j) \tag{4.10}$$

is an inexact subset.

Proof. It is trivial that

$$p_j \in S_i(\mathcal{A}(-i) \cup \mathcal{A}(+i, +j)) \tag{4.11}$$

Suppose that S is an element of (4.10) containing $p_{\sigma(j)}$. Then S does not contain p_j and, by (4.11), we have $p_i \notin S$. The latter guarantees that $p_{\sigma(i)} \in S$. Therefore, $p_{\sigma(i)}$ belongs to every element of (4.10) containing $p_{\sigma(j)}$ and

$$p_{\sigma(i)} \in S_{\sigma(j)}(\mathcal{A}(-i) \cup \mathcal{A}(+i, +j)).$$

As in the proof of Lemma 4.17, we construct a frame

$$(B \setminus \{p_i, p_{\sigma(j)}\}) \cup \{p_i', p_{\sigma(j)}'\}$$

which defines a new apartment of $\mathcal{G}_\delta(\Pi)$ containing (4.10). □

Proposition 4.34. *If $\mathcal{R}$ is a maximal inexact subsets of $\mathcal{A}$ then*

$$\mathcal{R} = \mathcal{A}(-i) \cup \mathcal{A}(+i, +j)$$

for certain i, j such that $j \neq i, \sigma(i)$.

Proof. Since $\mathcal{R}$ is inexact, we have $S_i(\mathcal{R}) \neq p_i$ for a certain number i. If $S_i(\mathcal{R})$ is empty then $\mathcal{R}$ is contained in the non-maximal inexact subset $\mathcal{A}(-i)$ which contradicts the assumption that the inexact subset $\mathcal{R}$ is maximal. Thus there exists $p_j \in S_i(\mathcal{R})$ such that $j \neq i$. An easy verification shows that

$$\mathcal{R} \subset \mathcal{A}(-i) \cup \mathcal{A}(+i, +j).$$

Since the latter subset is inexact and the inexact subset $\mathcal{R}$ is maximal, the inverse inclusion holds. $\qquad\square$

The complement subset corresponding to $\mathcal{A}(-i) \cup \mathcal{A}(+i, +j)$ is

$$\mathcal{A}(+i, -j) = \mathcal{A}(+i, +\sigma(j)).$$

By Proposition 4.34, every compliment subset of $\mathcal{A}$ coincides with certain $\mathcal{A}(+l, +m)$, $m \neq l, \sigma(l)$.

Lemma 4.22. *Subspaces $S, U \in \mathcal{A}$ are adjacent if and only if there are precisely*

$$\binom{n-2}{2}$$

distinct complement subsets of $\mathcal{A}$ containing S and U.

Proof. See the proof of Lemma 4.20 in the case when $k = n - 1$. $\qquad\square$

Remark 4.22. Suppose that $n = 4$. Then every complement subset of $\mathcal{A}$ consists of two adjacent elements. On the other hand, $\mathfrak{G}_\delta(\Pi)$ is a polar space of type D_4 and $\mathcal{A}$ is a frame of this space; by Proposition 4.31, in a polar space of type D_n every complement subset of a frame is a pair of collinear points.

Proposition 4.35. *Suppose that Π is a polar space of type D_n, $n \geq 4$, and $f : \mathcal{G}_\delta(\Pi) \to \mathcal{G}_\gamma(\Pi')$, $\delta, \gamma \in \{+, -\}$, is an apartments preserving mapping. Then f is adjacency preserving (two elements of $\mathcal{G}_\delta(\Pi)$ are adjacent if and only if their images are adjacent).*

Proof. Let $\mathcal{A}$ be an apartment of $\mathcal{G}_\delta(\Pi)$. As in the proof of Proposition 4.32, we establish that $\mathcal{R}$ is a complement subset of $\mathcal{A}$ if and only if $f(\mathcal{R})$ is a complement subset of the apartment $f(\mathcal{A})$. Then Lemma 4.22 gives the claim. $\qquad\square$

4.9.5 *Proof of Theorem 4.16*

By Subsection 4.6.1, we need to establish that every apartments preserving bijection is an isomorphism between the associated Grassmann graphs. For polar Grassmannians of index > 0 and half-spin Grassmannians this follows immediately from Propositions 4.32 and 4.35. For Grassmannians of index 0 we have the following.

Proposition 4.36. *For a mapping $f : P \rightarrow P'$ the following conditions are equivalent:*

(1) *f is apartments (frames) preserving,*
(2) *f is a collinearity preserving injection (two points are collinear if and only if their images are collinear).*

Proof. The implication $(2) \Longrightarrow (1)$ easy follows from the frame definition.

$(1) \Longrightarrow (2)$. Let $f : P \rightarrow P'$ be a mapping which sends frames of Π to frames of Π'. This mapping is injective. Since any two points are contained in a certain frame, it is sufficient to show that two points in a frame $B = \{p_1, \ldots, p_{2n}\}$ of Π are collinear if and only if their images are collinear. As in the proof of Propositions 4.32, we establish that a subset $X \subset B$ is complement if and only if $f(X)$ is a complement subset of the frame $f(B)$. In the case when Π and Π' are of type D_n, each complement subset of B is a pair of collinear points and we get the claim.

Suppose that our polar spaces are of type C_n. For each $i \in \{1, \ldots, 2n\}$ there exists a point p_i' such that

$$(B \setminus \{p_i\}) \cup \{p_i'\} \text{ and } (B \setminus \{p_{\sigma(i)}\}) \cup \{p_i'\}$$

are frames of Π (Lemma 4.18). Then

$$(f(B) \setminus \{f(p_i)\}) \cup \{f(p_i')\} \text{ and } (f(B) \setminus \{f(p_{\sigma(i)})\}) \cup \{f(p_i')\} \quad (4.12)$$

are frames of Π'. Since $f(B)$ is a frame, there are unique $u, v \in \{1, \ldots, 2n\}$ such that

$$f(p_i) \not\perp f(p_u) \text{ and } f(p_{\sigma(i)}) \not\perp f(p_v). \quad (4.13)$$

It is clear that $u \neq v$ and $f(p_i')$ is non-collinear with both $f(p_u), f(p_v)$. Thus there is no frame of Π' containing the subset

$$\{f(p_i'), f(p_u), f(p_v)\}.$$

We have

$$\{f(p_i), f(p_{\sigma(i)})\} = \{f(p_u), f(p_v)\}$$

(otherwise, one of the frames (4.12) contains $f(p_i'), f(p_u), f(p_v)$). Then (4.13) implies that $u = \sigma(i)$ and $v = i$. Therefore, $f(p_i) \not\perp f(p_{\sigma(i)})$ which guarantees that $f(p_i)$ is collinear with $f(p_j)$ if $j \neq \sigma(i)$. $\square$

4.9.6 *Embeddings*

Proposition 4.37. *Every apartments preserving mapping $f : P \to P'$ is an embedding of Π in Π'.*

Proof. By Proposition 4.36, f is a collinearity preserving injection. Let $p_1, p_2 \in P$ be distinct collinear points and $\{p_1, p_2, \ldots, p_{2n}\}$ be a frame of Π containing them. We suppose that

$$p_i \not\perp p_{i+n} \quad \forall\, i \in \{1, \ldots, n\}.$$

Then all $p_i' := f(p_i)$ form a frame of Π' and

$$p_i' \not\perp p_{i+n}' \quad \forall\, i \in \{1, \ldots, n\}.$$

The line $p_1 p_2$ is a maximal singular subspace of the generalized quadrangle

$$\{p_3, \ldots, p_n, p_{n+3}, \ldots, p_{2n}\}^\perp$$

(Lemma 4.3) and we have

$$p_1 p_2 = \{p_1, p_2, p_3, \ldots, p_n, p_{n+3}, \ldots, p_{2n}\}^\perp.$$

Similarly, we obtain that

$$p_1' p_2' = \{p_1', p_2', p_3', \ldots, p_n', p_{n+3}', \ldots, p_{2n}'\}^\perp.$$

Thus a point $p \in P$ is on the line $p_1 p_2$ if and only if $f(p)$ is on the line $p_1' p_2' = f(p_1) f(p_2)$. This means that f transfers lines to subsets of lines and distinct lines go to subsets of distinct lines. $\square$

Let $f : P \to P'$ be an apartments preserving mapping. By Proposition 4.37, it is an embedding of Π in Π'. For every singular subspace S of Π there is a frame such that S is spanned by a subset of this frame. This implies that $f(S)$ consists of mutually collinear points and

$$\dim\langle f(S)\rangle = \dim S.$$

So, for every $k \in \{1, \ldots, n-1\}$ we have the mapping

$$(f)_k : \mathcal{G}_k(\Pi) \to \mathcal{G}_k(\Pi')$$

$$S \to \langle f(S)\rangle.$$

An easy verification shows that this is an apartments preserving embedding of $\mathfrak{G}_k(\Pi)$ in $\mathfrak{G}_k(\Pi')$.

Similarly, we construct the apartments preserving embedding $(f)_\delta$ of $\mathfrak{G}_\delta(\Pi)$ in $\mathfrak{G}_\gamma(\Pi')$, $\delta, \gamma \in \{+, -\}$, if our polar spaces are of type D_n.

Theorem 4.17 (M. Pankov). *Let $1 \leq k \leq n - 1$. If $k \neq 1$ or the polar spaces are not of type D_4 then every apartments preserving mapping of $\mathcal{G}_k(\Pi)$ to $\mathcal{G}_k(\Pi')$ is the embedding of $\mathfrak{G}_k(\Pi)$ in $\mathfrak{G}_k(\Pi')$ induced by an apartments preserving embedding of Π in Π'. If our polar spaces are of type D_n and $n \geq 5$ then every apartments preserving mapping of $\mathcal{G}_\delta(\Pi)$ to $\mathcal{G}_\gamma(\Pi')$, $\delta, \gamma \in \{+, -\}$, is the embedding of $\mathfrak{G}_\delta(\Pi)$ in $\mathfrak{G}_\gamma(\Pi')$ induced by an apartments preserving embedding of Π in Π'.*

Remark 4.23. In [Pankov 2 (2007)] a such kind result was proved for symplectic Grassmannians; in [Pankov 3 (2007)] it was announced for all Grassmannians associated with polar spaces of type C_n.

Now suppose that our polar spaces are of type D_4. In this case, the associated half-spin Grassmann spaces are polar spaces of type D_4. Let f be an apartments preserving embedding of Π in $\mathfrak{G}_\delta(\Pi')$, $\delta \in \{+, -\}$. Then f induces an embedding of $\mathfrak{G}_1(\Pi)$ in $\mathfrak{G}_1(\Pi')$; by Corollary 4.4, this embedding is apartments preserving. Also f induces an apartments preserving embedding of one of the half-spin Grassmann spaces of Π in $\mathfrak{G}_{-\delta}(\Pi')$.

Theorem 4.18 (M. Pankov). *Suppose that the polar spaces are of type D_4. Then every apartments preserving mapping of $\mathcal{G}_1(\Pi)$ to $\mathcal{G}_1(\Pi')$ is the embedding of $\mathfrak{G}_1(\Pi)$ in $\mathfrak{G}_1(\Pi')$ induced by an apartments preserving embedding of Π in Π' or in one of the half-spin Grassmann spaces of Π'. Similarly, every apartments preserving mapping of $\mathcal{G}_\delta(\Pi)$ to $\mathcal{G}_\gamma(\Pi')$, $\delta, \gamma \in \{+, -\}$, is the embedding of $\mathfrak{G}_\delta(\Pi)$ in $\mathfrak{G}_\gamma(\Pi')$ induced by an apartments preserving embedding of Π in Π' or in $\mathfrak{G}_{-\gamma}(\Pi')$.*

4.9.7   *Proof of Theorems 4.17 and 4.18*

Let $1 \leq k \leq n - 1$ and $f : \mathcal{G}_k(\Pi) \to \mathcal{G}_k(\Pi')$ be an apartments preserving mapping. By Proposition 4.32, it is adjacency preserving (two elements of $\mathcal{G}_k(\Pi)$ are adjacent if and only if their images are adjacent). In the case when $k \leq n - 2$, the mapping f preserves the weak adjacency relation (two elements of $\mathcal{G}_k(\Pi)$ are weakly adjacent if and only if their images are weakly adjacent) if $k \neq 1$ or the polar spaces are not of type D_4.

1. If $k = n - 1$ then f maps lines to subsets of lines and induces a mapping

$$f_{n-2} : \mathcal{G}_{n-2}(\Pi) \to \mathcal{G}_{n-2}(\Pi').$$

This mapping is injective (otherwise there exists a line of $\mathfrak{G}_{n-1}(\Pi')$ containing the images of two distinct lines of $\mathfrak{G}_{n-1}(\Pi)$, this implies the existence of non-adjacent elements of $\mathcal{G}_{n-1}(\Pi)$ whose images are adjacent).

If $1 \le k \le n - 2$ and f preserves the weak adjacency relation then it transfers maximal cliques of $\Gamma_k^w(\Pi)$ to cliques of $\Gamma_k^w(\Pi')$. Big stars go to subsets of big stars (since any two elements of a top are adjacent and a big star contains non-adjacent elements). The image of every big star is contained in precisely one big star (the intersection of two distinct big stars is empty or consists of one element). Therefore, f induces a mapping

$$f_{k-1} : \mathcal{G}_{k-1}(\Pi) \to \mathcal{G}_{k-1}(\Pi').$$

It is injective (otherwise there exist weakly adjacent elements of $\mathcal{G}_k(\Pi')$ whose pre-images are not weakly adjacent).

It must be pointed out that the mapping f_{k-1} is not defined only in the case when $k = 1$ and our polar spaces are of type D_4.

Let B be a frame of Π and $\mathcal{A}$ be the associated apartment of $\mathcal{G}_k(\Pi)$. Denote by B' the frame of Π' corresponding to the apartment $f(\mathcal{A})$. As in the proof of Theorem 3.10 (Subsection 3.4.3), we show that f_{k-1} maps the apartment of $\mathcal{G}_{k-1}(\Pi)$ associated with B to the apartment of $\mathcal{G}_{k-1}(\Pi')$ defined by B'. Therefore, f_{k-1} is apartments preserving.

In the case when $n \ne 4$ or the polar spaces are of type C_4, we construct a sequence of apartments preserving mappings

$$f_i : \mathcal{G}_i(\Pi) \to \mathcal{G}_i(\Pi'), \quad i = k, \dots, 0,$$

such that $f_k = f$,

$$f_i([S\rangle_i) \subset [f_{i-1}(S)\rangle_i$$

for all $S \in \mathcal{G}_{i-1}(\Pi)$ and

$$f_{i-1}(\langle U]_{i-1}) \subset \langle f_i(U)]_{i-1}$$

for all $U \in \mathcal{G}_i(\Pi)$ if $i \ge 1$. An easy verification shows that each f_i, $i \ge 1$, is induced by f_0.

2. Now suppose that our polar spaces are of type D_4.

Consider the case $k = 2$. Since f_1 is apartments preserving, it is adjacency preserving and maximal singular subspaces of $\mathfrak{G}_1(\Pi)$ go to subsets of maximal singular subspaces of $\mathfrak{G}_1(\Pi')$. On the other hand, f_1 is induced by f; this implies that f_1 maps tops to subsets of tops. Then stars go to subsets of stars (for every star there is a top intersecting this star in a line and the intersection of two distinct tops contains at most one element). In the present case, two stars of $\mathfrak{G}_1(\Pi)$ are adjacent (Subsection 4.6.3) if and only if their intersection is a line. As in Subsection 4.6.3, we establish that

f_1 transfers big stars to subsets of big stars. The image of every big star is contained in precisely one big star and f_1 induces a mapping $f_0 : P \to P'$. It is not difficult to prove that f_0 is an apartments preserving embedding of Π in Π' which induces f_1 and f.

If $k = 3$ then f_2 is induced by an apartments preserving embedding of Π to Π' (it was established above). It is clear that this embedding also induces f.

Let $k = 1$. Then f is an adjacency preserving injection of $\mathcal{G}_1(\Pi)$ to $\mathcal{G}_1(\Pi')$ satisfying the condition (A) from Subsection 4.6.5. It transfers maximal singular subspaces of $\mathfrak{G}_1(\Pi)$ to subsets of maximal singular subspaces of $\mathfrak{G}_1(\Pi')$ and distinct maximal singular subspaces go to subsets of distinct maximal singular subspaces. The image of every maximal singular subspace of $\mathfrak{G}_1(\Pi)$ is contained in precisely one maximal singular subspace of $\mathfrak{G}_1(\Pi')$. If stars go to subsets of stars then, by the arguments given above, f is induced by an apartments preserving embedding of Π in Π'. In the case when the image of a certain star is contained in a top, we have the mapping $h : P \to \mathcal{G}_{-\gamma}(\Pi')$ (Subsection 4.6.5); it sends every line $L \in \mathcal{L}$ to a subset of the line of $\mathfrak{G}_{-\gamma}(\Pi')$ corresponding to $f(L)$ (Remark 4.12). The reader can show that h is an apartments preserving embedding of Π in $\mathfrak{G}_{-\gamma}(\Pi')$ which induces f.

3. Suppose that our polar spaces are of type D_n and g is an apartments preserving mapping of $\mathcal{G}_\delta(\Pi)$ to $\mathcal{G}_\gamma(\Pi')$, $\delta, \gamma \in \{+, -\}$. By Proposition 4.35, the mapping g is adjacency preserving (two elements of $\mathcal{G}_\delta(\Pi)$ are adjacent if and only if their images are adjacent). Then it transfers maximal singular subspaces of $\mathfrak{G}_\delta(\Pi)$ (stars and special subspaces) to subsets of maximal singular subspaces of $\mathfrak{G}_\gamma(\Pi')$. As above, distinct maximal singular subspaces go to subsets of distinct maximal singular subspaces.

There are the following four possibilities for the intersection of two distinct maximal singular subspaces: the empty set, a single point, a line, and a plane. For every plane there is an apartment intersecting this plane in a triangle. In the other cases, the intersections with apartments contain at most two elements. This means that g transfers every plane to a subset spanning a plane.

Show that the images of distinct planes are contained in distinct planes. This guarantees that g maps lines to subsets of lines.

Let $\mathcal{X}_1$ and $\mathcal{X}_2$ be planes of $\mathfrak{G}_\delta(\Pi)$ such that $g(\mathcal{X}_1)$ and $g(\mathcal{X}_2)$ are contained in a plane $\mathcal{X}'$ of $\mathfrak{G}_\gamma(\Pi')$. Then every element of $\mathcal{X}_1$ is adjacent with every element of $\mathcal{X}_2$ and there exists a maximal singular subspace of $\mathfrak{G}_\delta(\Pi)$ containing both these planes. This implies the existence of an apartment

$\mathcal{A} \subset \mathcal{G}_\delta(\Pi)$ which intersects each $\mathcal{X}_i$ in a triangle; but the apartment $g(\mathcal{A})$ intersects $\mathcal{X}'$ in a subset containing at most three points. This is possible only in the case when $\mathcal{X}_1 = \mathcal{X}_2$.

Similarly, we establish that distinct lines go to subsets of distinct lines. Therefore, g induces an injection of $\mathcal{G}_{n-3}(\Pi)$ to $\mathcal{G}_{n-3}(\Pi')$. This mapping is apartments preserving (standard verification); hence it is induced by an apartments preserving embedding of Π in Π' or in one of the half-spin Grassmann spaces of Π' (the second possibility can be realized only for $n = 4$). This embedding induces g.

Remark 4.24. In the case when $n = 4$, the half-spin Grassmann spaces $\mathfrak{G}_\delta(\Pi)$ and $\mathfrak{G}_\gamma(\Pi')$ are polar spaces of type D_4. By Corollary 4.4 and Proposition 4.37, g is an embedding of $\mathfrak{G}_\delta(\Pi)$ in $\mathfrak{G}_\gamma(\Pi')$. This implies that it transfers lines to subsets of lines and distinct lines go to subsets of distinct lines.

Bibliography

Abramenko, P. (1996). *Twin buildings and applications to S-arithmetic groups*, Lecture Notes in Mathematics **1641** (Springer-Verlag, Berlin).

Abramenko, P. and Van Maldeghem, H. (2000). On opposition in spherical buildings and twin buildings, *Ann. Comb.* **4**, 2, pp. 125–137.

Abramenko, P. and Van Maldeghem, H. (2008). Intersections of apartments, Preprint arXiv:0805.4442, to appear in *J. Combin. Theory Ser. A*.

Artin, E. (1957). *Geometric Algebra* (Interscience Publishers, New York-London).

Baer, R. (1952). *Linear Algebra and Projective Geometry* (Academic Press, New York).

Blok, R. J. and Brouwer, A. E. (1998). Spanning point-line geometries in buildings of spherical type, *J. Geom.* **62**, 1-2, pp. 26–35.

Blunck, A. and Havlicek, H. (2005). On bijections that preserve complementarity of subspaces, *Discrete Math.* **301**, 1, pp. 46–56.

Borovik, A. V., Gelfand, I. M. and White, N. (2003). *Coxeter Matroids*, Progress in Mathematics **216** (Birkhäuser, Boston).

Bourbaki, N. (1968). *Éléments de mathématique. Groupes et algèbres de Lie*, Chapters IV–VI, Actualités Scientifiques et Industrielles **1337** (Hermann, Paris).

Buekenhout, F. and Cameron, P. (1995). Projective and affine geometry over division rings, in F. Buekenhout (ed.), *Handbook of Incidence Geometry: Buildings and Foundations* (Elsevier, Amsterdam), pp. 27–62.

Buekenhout, F. and Shult, E. (1974). On the foundations of polar geometry, *Geom. Dedicata* **3**, 2, pp. 155–170.

Buekenhout, F. (1990). On the foundations of polar geometry II, *Geom. Dedicata* **33**, 1, pp. 21–26.

Brown, K. S. (1989). *Buildings* (Springer-Verlag, New York).

Cameron, P. (1982). Dual polar spaces, *Geom. Dedicata* **12**, 1, pp. 75–85.

Cameron, P. (1991). *Projective and polar spaces*, QMW Maths Notes **13** (Queen Mary and Westfield College, London).

Ceccherini, P. V. (1967). Collineazioni e semicollineazioni tra spazi affini o proiettivi, *Rend. Mat. e Appl.* **26**, pp. 309–348.

Chow, W. L. (1949). On the geometry of algebraic homogeneous spaces, *Ann. of*

Math. **50**, 1, pp. 32–67.

Cohen, A. M. (1995). Point-line spaces related to buildings, in F. Buekenhout (ed.), *Handbook of Incidence Geometry: Buildings and Foundations* (Elsevier, Amsterdam), pp. 647–737.

Cooperstein, B. N. and Shult, E.E. (1997). Frames and bases of Lie incidence geometries, *J. Geom.* **60**, 1-2, pp. 17–46.

Cooperstein, B. N. (2003). Generation of embeddable incidence geometries: a survey. Topics in diagram geometry, *Quad. Mat.* **12** (Dept. Math., Seconda Univ. Napoli, Caserta), pp. 29–57.

Cooperstein, B. N., Kasikova, A. and Shult, E. E. (2005). Witt-type Theorems for Grassmannians and Lie Incidence Geometries, *Adv. Geom.* **5**, 1, pp. 15–36.

Cooperstein, B. N. (2005). Classical subspaces of symplectic Grassmannians, *Bull. Belg. Math. Soc. Simon Stevin* **12**, 5, pp. 719–725.

Cooperstein, B. N. (2007). Symplectic subspaces of symplectic Grassmannians, *European. J. Combin.* **28**, 5, pp. 1442–1454.

Dieudonné, J. (1951). On the automorphisms of the classical groups, *Mem. Amer. Math. Soc.* **2**.

Dieudonné, J. (1951). Algebraic homogeneous spaces over field of characteristic two, *Proc. Amer. Math. Soc.* **2**, pp. 295–304.

Dieudonné, J. (1971). *La Géométrie des Groupes Classiques*, Ergebnisse der Mathematik und ihrer Grenzgebiete **5**, 3d edn. (Springer-Verlag, Berlin-New York).

Garrett, P. (1997). *Buildings and Classical groups* (Chapman & Hall, London).

Faure, C. A. and Frölicher, A. (1994). Morphisms of projective geometries and semilinear maps, *Geom. Dedicata* **53**, 3, pp. 237–262.

Faure, C. A. (2002). An elementary proof of the Fundamental Theorem of Projective Geometry, *Geom. Dedicata* **90**, 1, pp. 145–151.

Fillmore, P. A. and Longstaff, W. E. (1984). On isomorphisms of lattices of closed subspaces, *Canad. J. Math.* **36**, 5, pp. 820–829.

Havlicek, H. (1994). A generalization of Brauner's theorem on linear mappings, *Mitt. Math. Sem. Giessen* **215**, pp. 27–41.

Havlicek, H. (1995). On Isomorphisms of Grassmann Spaces, *Mitt. Math. Ges. Hamburg* **14**, pp. 117–120.

Havlicek, H. (1999). Chow's theorem for linear spaces, *Discrete Math.* **208/209**, pp. 319–324.

Havlicek, H. and Pankov, M. (2005). Transformations on the product of Grassmann spaces, *Demonstratio Math.* **38**, 3, pp. 675–688.

Huang, W.-l. and Havlicek, H. (2008). Diameter preserving surjections in the geometry of matrices, *Lin. Algebra Appl.* **429**, 1, pp. 376–386.

Huang, W.-l. and Kreuzer, A. (1995). Basis preserving maps of linear spaces, *Arch. Math. (Basel)* **64**, 6, pp. 530–533.

Huang, W.-l. (1998). Adjacency preserving transformations of Grassmann spaces, *Abh. Math. Sem. Univ. Hamburg* **68**, pp. 65–77.

Huang, W.-l. (2000). Adjacency preserving mappings of invariant subspaces of a null system, *Proc. Amer. Math. Soc.* **128**, 8, pp. 2451–2455.

Huang, W.-l. (2001). Characterization of the transformation group of the space of a null system, *Results Math.* **40**, 1-4, pp. 226–232.

Humphreys, J. E. (1975). *Linear algebraic groups*, Graduate Texts in Mathematics **21** (Springer-Verlag, New York-Heidelberg).

Kakutani, S. and Mackey, G.W (1946). Ring and lattice characterizations of complex Hilbert space, *Bull. Amer. Math. Soc.* **52**, 8, pp. 727–733.

Karzel, H., Sörensen, K. and Windelberg, D. (1973). *Einführung in die Geometrie*, Uni-Taschenbücher **184** (Vandenhoeck & Ruprecht, Göttingen).

Kreuzer, A. (1996). On the definition of isomorphisms of linear spaces, *Geom. Dedicata* **61**, 3, pp. 279–283.

Kreuzer, A. (1998). On isomorphisms of Grassmann spaces, *Aequationes Math.* **56**, 3, pp. 243–250.

Kwiatkowski, M. and Pankov, M. (2009). Opposite relation on dual polar spaces and half-spin Grassmann spaces, *Results Math.* **54**, 3-4, pp. 301–308.

Lim, M. H. (2010). Surjections on Grassmannians preserving pairs of elements with bounded distance, *Linear Algebra Appl.* **432**, 7, pp. 1703–1707.

Mackey, G. W. (1942). Isomorphisms of normed linear spaces, *Ann. of Math.* **43**, 2, pp. 244–260.

Pankov, M. (2002). Transformations of grassmannians and automorphisms of classical groups, *J. Geom.* **75**, 1-2, pp. 132–150.

Pankov, M. (2004). Chow's theorem and projective polarities, *Geom. Dedicata* **107**, pp. 17–24.

Pankov, M. (2004). Transformations of Grassmannians preserving the class of base sets, *J. Geom.* **79**, 1-2, pp. 169–176.

Pankov, M. (2004). A characterization of geometrical mappings of grassmann spaces, *Results Math.* **45**, 3-4, pp. 319–327.

Pankov, M. (2004). Mappings of the sets of invariant subspaces of null systems, *Beiträge Algebra Geom.* **45**, 2, pp. 389–399.

Pankov, M., Prażmowski, K. and Żynel, M. (2005). Transformations preserving adjacency and base subsets of spine spaces, *Abh. Math. Sem. Univ. Hamb.* **75**, pp. 21–50.

Pankov, M. (2005). On the geometry of linear involutions, *Adv. Geom.* **5**, 3, pp. 455–467.

Pankov, M., Prażmowski, K. and Żynel, M. (2006). Geometry of polar grassmann spaces, *Demonstratio Math.* **39**, 3, pp. 625–637.

Pankov, M. (2006). Geometrical mappings of Grassmannians of linear spaces satisfying the exchange axiom, *Aequationes Math.* **72**, 3, pp. 254–268.

Pankov, M. (2007). Base subsets of Grassmannians: infinite-dimensional case, *European. J. Combin.* **28**, 1, pp. 26–32.

Pankov, M. (2007). Base subsets of symplectic Grassmannians, *J. Algebraic Combin.* **26**, 2, pp. 143–159.

Pankov, M. (2007). Base subsets of polar Grassmannians, *J. Combin. Theory Ser. A* **114**, 8, pp. 1394–1406.

Pasini, A. (1994). *Diagram geometries*, Oxford Science Publications (Clarendon Press, Oxford University Press, New York).

Prażmowski, K. (2001). On a construction of affine Grassmannians and spine spaces, *J. Geom.* **72**, 1-2, pp. 172–187.

Prażmowski, K. and Żynel, M. (2002). Automorphisms of spine spaces, *Abh. Math. Sem. Univ. Hamb.* **72**, pp. 59–77.

Prażmowski, K. and Żynel, M. (2009). Segre subproduct, its geometry, automorphisms and examples, *J. Geom.* **92**, 1-2, pp. 117-142.

Rickart, C. E. (1950). Isomorphic groups of linear transformations, *Amer. J. Math.* **72**, pp. 451-464.

Ronan, M. (1989). *Lectures on buildings*, Perspectives in Mathematics **7** (Academic Press, Boston).

Rudin, W. (1973). *Function Analysis*, McGraw-Hill Series in Higher Mathematics (McGraw-Hill, New York-Düsseldorf-Johannesburg).

Scharlau, R. (1995). Buildings, in F. Buekenhout (ed.), *Handbook of Incidence Geometry: Buildings and Foundations* (Elsevier, Amsterdam), pp. 477–646.

Taylor, D. E. (1992). *Geometry of the classical groups*, Sigma Series in Pure Mathematics **9** (Heldermann Verlag, Berlin).

Teirlinck, L. (1980). On projective and affine hyperplanes, *J. Combin. Theory Ser. A* **28**, 3, pp. 290–306.

Tits, J. (1974). *Buildings of spherical type and finite BN-pairs*, Lecture Notes in Mathematics **386** (Springer-Verlag, Berlin-New York).

Veldkamp, F. D. (1959/1960). Polar geometry I – V, *Nederl. Akad. Wet. Proc. Ser. A* 62, pp. 512–551, Nederl. Akad. Wet. Proc. Ser. A 63, pp. 207–212.

Wan, Z.-X. (1996). *Geometry of Matrices* (World Scientific, Singapore).

Zick, W. (1983). Lineare Abbildungen in reellen Vektorräumen, *Math. Semesterber* **30**, 2, pp. 167–170.

Żynel, M. (2000). Subspaces and embeddings of spaces of pensils, Preprint, University in Białystok.

Index